Scientific Cultivation of Mushroom

Scientific Cultivation of Mushroom

Dr. Nikulsinh M. Chauhan
M.Sc. Agri., Ph.D.
Programme Coordinator, KVK, NAU, Vyara, Gujarat

Dr. Nirav Gajre
Farm Manager

Viral P. Prajapati
Senior Research Fellow

BIOTECH

2013

BIOTECH BOOKS®

ISBN 978-81-7622-268-6

Published by: **BIOTECH BOOKS®**
4762-63/23, Ansari Road, Darya Ganj,
New Delhi - 110 002
Phone: +91-011-23262132
E-mail: biotechbooks@yahoo.co.in

Printed at: **Chawla Offset Printers**
Delhi - 110 052

PRINTED IN INDIA

Preface

"Vision with action can change the world"

Joel A. Barker

This book is addressed to those who love mushrooms, who love the whole rich world of mushrooms in the same way that many love the flowers of the field and the birds of the air.

Mushrooms are a very important crop and are grown commercially in many countries. China grows more types of mushrooms than any other country and is the largest overall producer accounting for some 32 per cent . Worldwide, the button mushroom or white mushroom is probably the most widely grown. It is eaten both fresh and processed, especially canned. Modern production methods are highly mechanized, requiring detailed knowledge and a high level of management skill for successful and continuous cropping. Inevitably investment and production costs are high and returns have not kept pace with inflation, though the crop is still one of the most valuable in many of the countries where it is grown.

The cultivation of Mushrooms is a carefully controlled biological system; however contamination with microorganisms, which are in ways, is inevitable. In India majority of the mushroom holdings are lacking adequate compost preparation, pasteurization and proper environmental control facilities, which lead to the development of various diseases and pests sufficiently to a level to cause considerable yield loss. It is therefore very important for the mushroom growers that they should know the importance of diseases and competitors and should understand the importance of hygiene to grow mushrooms successfully and profitably. I would like to advise the mushroom growers to pay maximum attention to prepare compost/ substrate of optimum quality and maintain highest level of hygiene to avoid these problems.

The aim of this book is provide information on the problems of the crop and the best ways to overcome them to remain in business. In this respect, all the chapters have more informative for growers. The present book *"Scientific Cultivation of Mashroom"* is an attempt to provide graduate, post-graduate students and progressive farmers of India with a tool, which will assist them in supplementing their efforts for growing mushroom. Discussion with the farmers has brought out the fact that a handy good quality Mushroom for cultivation was long demanded. Therefore, we endeavored to compile comprehensive information for mushroom cultivation.

This is a compilation work and embodies a fairly comprehensive and crisp treatment of the fundamental facts and aspects of mushroom to prepare farmers to achieve success in growing of mushroom.

The book deals with the landmark in the field of Mushroom growing region, brief introduction about mushroom, their biology and diagnosis, classification and management of disease and pest.

We wish that students, farmers and various readers who relish working in the field of mushroom growing will find the format of book, level and quantity of information contained in the book to be appropriate for easier learning and understanding. The wide spectrum information in various chapters with the addition of the terms related to chapter and concept statements is presented in very concise manner. It is expected that the mushroom growers will find the format stimulating and conducive to learning and preparing themselves with confidence. Your valuable comments and suggestion are most welcome to improve it further.

N.M. Chauhan

N. Gajre

V.P. Prajapati

Contents

Chapter 1
Introduction

Mushrooms have been described as "Food for God" in classic writings of "Vedas". The earliest word in Sanskrit appeared to be *Ksumpa,* in Hindi as *Kumbhi* in Gujarati it is known as *Biladino Top*. Chinese people were the first to do the artificial cultivation of the mushroom about thousand years ago systematic study of mushroom fungi is only 250 years old real commercial cultivation of button mushroom started by Europeans during 16th and 17th century. Mushroom cultivation in India began in the year 1962 in Himachal Pradesh at Solan. The first improvised small mushroom farm was established by the Government of India at Solan in 1965. Since then mushroom production in India has increased from 7000 tons in the year 1983 to 70,000 tons in the year 2004. This production is quite low as compared to other mushroom growing countries of the world.

Mushrooms are not group of edible fungi they belong to the plant nor to the animal do they belong to separate group of organism call fungi (Mycota). They lack chlorophyll so cannot prepare their own food depend on the outside source for their nutritional requirement. Mushrooms are macro fruiting bodies or reproductive structures emanating from the mycelium with a distinctive fruiting body which can be either epigeous or hypogeous and large enough to be seen by naked eye and to be picked by hand. Not all mushrooms are edible, but some are highly poisonous, while edible fleshy fungi are called mushrooms, poisonous ones are termed 'toadstools'. It has been estimated that out of 10,000 species of fleshy fungi about half of them are edible and as many as 100 species are highly poisonous.

Chapter 2
Biology of Mushrooms

Mushrooms belong to the kingdom of Fungi, a group very distinct from plants, animals and bacteria. Fungi lack the most important feature of plants: the ability to use energy from the sun directly through chlorophyll. Thus, fungi depend on other organisms for food, absorbing nutrients from the organic material in which they live. The living body of the fungus is mycelium made out of a tiny web of threads (or filaments) called hyphae. Under specific conditions, sexually compatible hyphae will fuse and start to form spores. The larger spore-producing structures (bigger than about 1 mm) are called mushrooms. In nature this is the most striking part of the organism, but in fact it is just the fruiting body and the major part of the living organism is found under the ground or inside the wood.

Life Cycle of Fungi

Fungi multiply by producing millions and millions of

spores. When a spore settles in a suitable environment, it can germinate and branch to form a mycelium. When two sexually compatible mycelia meet, they may fuse to form a so-called secondary mycelium, which is capable of forming fruiting bodies.

Mycelial Growth and Spawn

In the practice of edible mushroom cultivation no use is made of spo-res. Their small size makes them difficult to handle and their genetic characteristics may differ from those of their parent. Moreover, it takes some time for mushroom spores to germinate, whereas other fungi such as green moulds germinate and spread much faster.

The desired mushroom must be able to colonise the substrate before other fungi or bacteria do so. To achieve

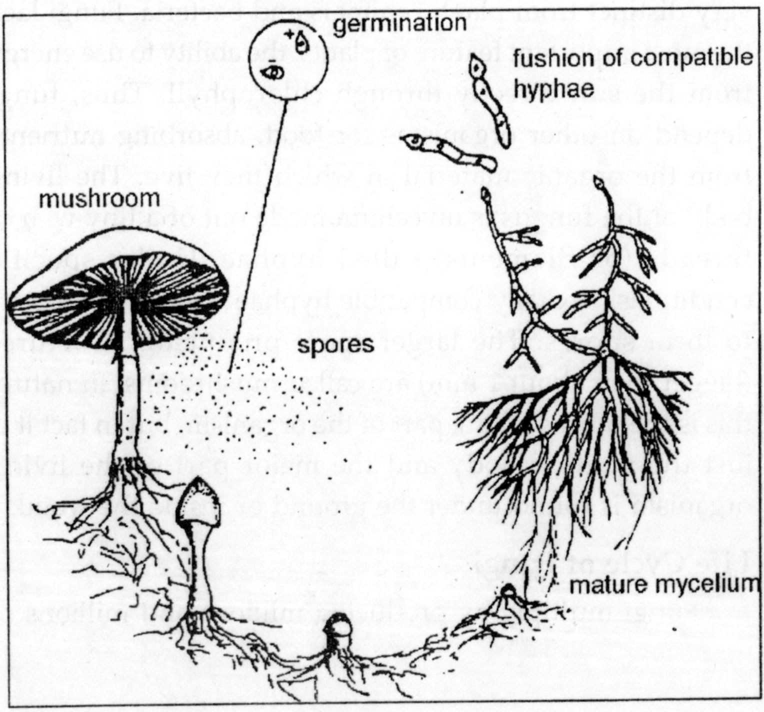

this, pre-grown mycelium (free of any contaminants) of the mushroom is inoculated on a sterile substrate. This material is referred to as spawn. Using spawn will give the cultivated mushroom an advantage in growth over other fungi.

Spawn Run

The mycelium will colonise the substrate and use the available nutri-ents. This is commonly referred to as the spawn run. When some nutrients run out, or when the weather changes, the mycelium will reach a different phase: the reproductive stage. A temperature of about 25°C is optimal for the spawn run of most species. The environment can also enhance the growth of the desired mycelium: a high CO_2 concentration is favourable for mycelial growth (but not for cropping). After having colonised the substrate, the mycelium is capable of producing fruiting bodies. The number and quality of the fruiting bodies will depend on the environment.

Chapter 3
The Potential of the Mushroom Industry in India

India is not a major producer of any of the mushroom varieties, but it does cultivate mushrooms and has great potential as an important producer in the future. From a production standpoint, the white button mushroom has the highest growth rate and potential for production. However, the cultivation of oyster mushrooms has been more common since the end of the last century, when the infrastructure of oyster mushroom growing was much improved and therefore capital requirements went way down in comparison with the requirements for white button mushroom cultivation.

Though India's present share in the world production and trade of oyster mushroom is meager, being only an estimated 2,000 tons, the potential for the future is rated as high for a variety of reasons. India has a very large availability of various types of raw substrate material such as wheat straw, paddy straw, bagasse, chicken manure, gypsum, tea waste, de-oiled cakes and so forth in almost all the regions and these materials are relatively inexpensive when compared with international prices. In 2001-2002, the production of wheat and paddy in India was estimated to be 73.53-90.75 million tons respectively. Although the residue straws are commonly used as fodder, almost 50 per cent of the crop residues are still potentially available for the growing of mushrooms.

India has large number of agro-climatic regions that offer congenial climatic conditions for mushroom cultivation. India also has a good combination of both the technical and non-technical manpower needed to operate and manage the mushroom growing operations. The supply and demand gap in the world trade of mushrooms and the shrinkage of production in countries like Taiwan and South Korea due to high labor costs would result in better market prices for Indian mushroom producers. The costs of building materials and other inputs related to construction costs are much lower in India than in many other countries. This keeps the investment cost per unit weight of mushroom produced more advantageous in India.

India is also developing its infrastructure rapidly and therefore enjoys a large and well-organized distribution network that facilitates the marketing of products in order to meet domestic consumer demands. From a dietary standpoint mushrooms are a particularly favorable food in vegetarian-predominant India. With a domestic population of more than one billion, India itself is a large market for mushrooms. The per capita consumption of mushrooms in India is currently only about 25g per year although there has been a steady increase in the consumption of exotic mushrooms including oyster mushrooms in addition to the use of regular button mushrooms. This increase can be seen as a highly encouraging sign coming from the potential mushroom consumers in India. Cultivated mushrooms are available today in all common vegetable shops, grocery stores, department stores in both small and big towns in India. One final reason for optimism concerning India's potential as a major mushroom producer is its strategic geographical location with respect to exportation, making it convenient to export oyster mushrooms to the Middle East, Europe, the United States, Africa, and Southeast Asia.

Chapter 4

Benefits of Mushroom Growing

There are many remarkable ecological advantages in the cultivation of edible fungi. One major advantage is the efficient re-integration of agricultural residues such as horse and chicken manure, cereal straw, bagasse and others. The spent mushroom substrate can then be used either as animal feed or as compost for application in farm fields.

The cost of oyster mushroom cultivation varies according to regions and the specific type of cultivation, but generally, the growing of oyster mushrooms is less expensive than that of other cash crops. The major reason for this is it requires little space and inexpensive raw materials. Oyster mushroom cultivation is economically efficient for the farmers of other crops, who do not have to buy the raw materials for substrate and can use low cost structures for mushroom cultivation on seasonal basis. Table 1 would provide a view on the cost-benefit relationships of oyster mushroom cultivation in India.

Table 1: Cost Benefits Relationship of Oyster Mushroom Cultivation in India

Growing Methods	*Production Cost (USD)*	*Yearly Production (tons)*	*Price Per kg* (USD)*	*Value of Sales (USD)*	*Earning Rate*
Traditional hut growing with purchased raw materials	16,383.06	27.37	1.1	30,107	45.58 per cent
Traditional hut growing with their own raw materials	12,364.78	27.37	1.1	30,107	58.94 per cent
Seasonal growing with purchased raw materials	6,832.34	11.4	0.95	10,830	36.91 per cent
Seasonal growing with their own raw materials	5,156.81	11.4	0.95	10,830	52.38 per cent
Growing in their own constructed houses	16,392.23	27.37	1.1	30,107	45.55 per cent

* Average yearly price realization for kg of Fresh Mushrooms.

As the table depicts, there are two most likely situations. Some growers are growing mushrooms with purchased raw materials, while others are growing mushrooms with their own raw materials. If the substrate materials are from the owner's own fields, this produces maximum profits. To obtain the maximum benefit mushroom growers should be farmers of other crops or young farmers in rural areas. The Indian agencies involved in the promotion of oyster mushroom growing are using this information to promote self-employment among rural youth. This particular aspect holds good for all developing nations in which rural youth are migrating to industrialized cities in search of employment.

4.1 Mushroom Nutritional and Medicinal Properties

Edible mushrooms provide high quality of protein that can be produced with greater biological efficiency than animal protein. They are rich in fiber, minerals and vitamins, and have low crude fat content, with a high proportion of polyunsaturated fatty acids (72 to 85 per cent) relative to total content of fatty acids. These properties are major contributing factors to the traditional recognition of mushrooms as "healthy" foods.

A large number of mushroom species are not only edible and nutritious but also possess tonic and medicinal qualities. However, some mushrooms are lethally poisonous, and one should eat mushrooms only if one knows their names and their properties with considerable precision.

In the past, the mushroom industry concentrated mainly on the production of fresh, canned and dried mushrooms for food. Thus, the industry had only one leg.

In the present era, high-pressure work demands are causing greater stress to the human body, and resulting in the weakening of the human immune system. A variety of proprietary products based on mushroom nutriceuticals and mushroom pharmaceuticals have already been produced and marketed. This trend is expected to increase with wider consumer satisfaction and acceptability. This is the second leg of the industry. These two legs/segments of the mushroom-based industry will not compete but will complement each other.

4.2 Nutritional Value of Mushrooms

The greatest difficulty in feeding man is to supply a sufficient quantity of the body-building material–protein. The other three nutritional categories are: the source of energy food– carbohydrates and fats; accessory food factors– vitamins; and inorganic compounds which are indispensable to good health. Of course, water, too, is essential.

The moisture content of fresh mushrooms varies within the range of 70 - 95 per cent depending upon the harvest time and environmental conditions, whereas it is about 10 - 13 per cent in dried mushrooms. The protein content of the cultivated species ranges from 1.75 to 5.9 per cent of their fresh weight. It has been estimated that an average value of 3.5 to 4.0 per cent would be more representative. This means that the protein content of edible mushrooms in general, is about twice that of onion (1.4 per cent) and cabbage (1.4 per cent), and four times and 12 times those of oranges (1.0 per cent) and apples (0.3 per cent), respectively. In comparison, the protein content of common meats is as follows: pork, 9-16 per cent; beef, 12-20 per cent; chicken, 18-20 per cent; fish, 18 -20 per cent; and milk, 2.9-

3.3 per cent. On a dry weight basis, mushrooms normally contain 19 -35 per cent protein, as compared to 7.3 per cent in rice, 12.7 per cent in wheat, 38.1 per cent in soybean, and 9.4 per cent in corn. Therefore, in terms of the amount of crude protein, mushrooms rank below animal meats, but well above most other foods, including milk, which is an animal product (Chang and Miles 1989). Furthermore, mushroom protein contains all the nine essential amino acids required by man.

Quantitative data relating to the nutritive value of mushrooms is sparse. In the absence of feeding trials, alternative methods have been used to determine or predict the nutritional value of foods based on their content of essential amino acids (Crimson and Sands 1978).The Essential Amino Acid Index (EA. Index) rates dietary protein in terms of an essential amino acid pattern based on known adult human dietary requirements. The Amino Acid Score (Chemical Score) is the amount of the most limiting amino acid in the food protein expressed as a percentage of the same amino acid present in the reference protein. In an attempt to resolve the difficulties inherent in comparisons between those mushrooms containing small amounts of high quality protein with those containing larger amounts of a protein of lesser nutritional quality, Crisan and Sands (1978) proposed the use of a Nutritional Index calculated as:

$$\text{Nutritional Index} = \frac{(\text{EAA index x percentage protein})}{100}$$

The EAA Indices, the Amino Acid Scores, and Nutritional Indices for various mushrooms and other foods, are reported by Crisan and Sands (1978). EAA Indices and

Amino Acid Scores of the most nutritive mushrooms (highest values) rank in potential nutritive value with those of meat and milk, and are significantly higher than those for most legumes and vegetables. The least nutritive mushrooms rank appreciably lower but are still comparable to some of our common vegetables.

In addition to their good proteins, mushrooms are a relatively good source of the following individual nutrients: fat, phosphorus, iron, and vitamins including thiamine, riboflavin, ascorbic acid, ergosterine and niacin. They are low in calories, carbohydrates and calcium. It has also been reported that a total lipid content varying between 0.6 and 3.1 per cent of the dry weight, is found in the commonly cultivated mushrooms. At least 72 per cent of the total fatty acids are found to be unsaturated in all the four tested mushrooms (Huang, *et al.*, 1985). It should be noted that unsaturated fatty acids are essential and significant in our diet and to our health.

In recent years, there has been a trend toward discovering ways of treating mushrooms so as to give them added value. For example, Wermer and Beelman (2002) have reported on growing mushrooms enriched in selenium. By adding sodium selenite to compost over a range of 30-300 parts per million, they found that the mushrooms increasingly absorbed selenium according to the amount in the compost, so that it is possible to grow mushrooms containing a desired concentration. Selenium is an essential micronutrient that has generated much recent interest in nutritional and medical research–and, more recently, within the food industry. Selenium has numerous physiological functions, but is best known as necessary cofactor for the glutathione peroxidase enzyme system. This system is responsible for removing free radicals from the body, thus reducing oxidative damage.

The desirability of a food product does not necessarily bear any correlation to its nutritional value. However, its appearance, taste, and aroma, sometimes can stimulate one's appetite (preference). In addition to nutritional value, mushrooms have some unique colour, taste, aroma, and texture characteristics, which attract their consumption by humans.

4.3 Medicinal Properties of Mushrooms

The second major attribute of mushrooms, their medicinal properties, has long been recognised in China, Korea, and Japan. There has been a great upsurge in activities related to the use of mushroom products for medicinal purposes in recent years. In 2001, the figure of US$9-10 billion was cited as representing the value of medicinal mushroom products, including tonics and medicines. The application of modern analytical techniques can be used to establish a scientific basis for the empirical observations that have been made centuries before. According to Chang and Buswell (1996), the term *"mushroom nutriceutica F* is used for a new class of new compounds extractable from either the mycelium, or the fruiting body of the mushroom. Mushroom nutriceuticals may possess both nutritional and medicinal properties.

Due to present day high pressured work demands resulting in great stress to the human body and causing a weakening of the human immune system, there are now many new diseases. These have developed as a consequence of lower natural body resistance. There is some evidence that the beneficial treatment of these diseases can be obtained by consumption of mushrooms as a functional food, or through the use of extracted biologically active compounds as a dietary supplement, in order to enhance immune response of the human body, thereby increasing resistance to disease and, in some cases, causing regression of a diseased state. Differing from most pharmaceuticals, these biologically active compounds extracted from medicinal mushrooms have extraordinarily low toxicity, even at high doses. Long viewed as tonics, now it has been known that they can profoundly improve the quality of human health.

Mushrooms produce several biologically active compounds that are usually associated with the cell wall. Most notably, a group of polysaccharides comprising high molecular weight sugar polymers has been reported to contribute to their immune enhancing and tumour retarding effects. It has been reported that the anti-tumour and anti-cancer effects of the polysaccharides are based on the enhancement of the body's immune systems, including activated macrophages, natural killer cells, cytotoxic T cells, and their secretory products, such as the tumour necrosis factor, reactive nitrogen and oxygen intermediates, and interleukins, rather than direct cytocidal effects. It should be noted that immune responses are complex reactions involving several types of cells, such as macrophages and lymphocytes. The killing mediated by cytotoxic T lymphocytes and natural killer cells, represents an important mechanism in immune defence against tumours, virus-infected cells, parasites and other foreign invaders. Another group of medicinal compounds found in *Ganoderma* spp. are triterpenoids, steroid-like compounds, which contribute cytotoxic, hepatoprotective, and hypolipidmic influence on platelet aggregation, inhibition of angiotensine-converting enzyme, and inhibition of histamine release (Lindequist 1995). Lectins, another group of mushroom bioactive compounds, are proteins or glycoproteins with specific binding sites for sugars, which are not antibodies or enzymes. These have become useful tools in structural studies of the cell surface, oligosaccharides and/or carbohydrate moieties of glycoproteins. Some lectins have been shown to have anti-tumour and immunomodulatory activities (Wang *et al.*, 1996). Other lectins preferentially agglutinate mammalian cells that have been transformed by oncogenic viruses or by chemical carcinogens, as well as spontaneously transformed cells. These and related findings indicate that studies with lectins may lead to a better understanding of cancer. Moreover, some lectins may be used to inhibit the growth of malignant cells. A novel single-chained ribosome-inactivating protein (RIP) was recently isolated from fruit bodies of the edible mushroom, *V. volvacea* (Yao et al. 1998). The mushroom RIP, designated volvarin, exhibited a potent inhibitory action on protein synthesis in the rabbit reticulocyte lysate system. Like most plant RIPs, volvarin acted as an N-glycosidase that depurinated rRNA from rabbit reticulocyte lysate, releasing a characteristic RNA fragment after treatment with aniline. It also exerted a strong abortifacient effect in mice.

Furthermore, the aqueous extracts of *Pleurotus sajor-caju* (Tam *et al.*, 1986) and *Volvariella volvacea* (Chiu *et al.*, 1996) have been reported to produce a hypotensive effects in normotensive rats. Feeding powdered maitake *(Grifola frondosa)* mushrooms to spontaneous hypertensive rats resulted in a lowering of the blood pressure (Kyoko et al. 1988). It has also been reported that dried powder of another two edible mushrooms, *Auricularia auricula* and *Tremella fuciformis,* after being fed to the rats, has demonstrated to be effective in lowering both the serum total cholesterol and the low density lipoprotein (LDL) cholesterol levels (Cheung 1996). Since the mushrooms did not affect the concentration of serum high density lipoprotein (HDL) cholesterol, the reduction of serum total cholesterol by the mushroom diets is believed to be attributable to the fall in LDL cholesterol. It should be noted that LDL is "bad" cholesterol and HDL is "good" cholesterol.

4.4 Mushroom Nutriceuticals

There has been a recent upsurge of interest in mushrooms not only as health vegetables (food) but also as a source of biological active compounds of medicinal value, including use as complementary medicine/dietary supplements for anticancer, antiviral, immunopotentiating, hypocholesterolaemic and hepatoprotective agents. This new class of compounds termed 'mushroom nutriceuticals' (Chang and Buswell 1996), are extractable from either the fungal mycelium or fruiting body and represent an important component of the expanding mushroom biotechnology industry.

Of the 14,000-15,000 species of so-called mushrooms in the world, around 400 have known medicinal properties. However, it has been estimated that there are about 1,800 species of mushrooms with the potential of medicinal properties. Both these mushrooms and their root-like structure (called mycelium) produce several medicinal or nutriceutical (general immune enhancing) compounds, central of which are the polysaccharides (high molecular weight strings of sugars), triterpenes, and immunomodulatory proteins. Although virtually all mushrooms and many foods have polysaccharides in their cell walls, certain mushroom species have been found to contain polysaccharides which are particularly effective in retarding the progress of various cancers and other diseases, and in alleviating the side effects of chemotherapy and radiation treatment (through cell-level regenerative effects). There are now many studies in Asia, particularly in China and Japan, documenting life span increases of cancer patients undergoing conventional cancer treatment plus mushroom extract consumption or injection (Mizuno *et al.*, 1995; Liu 1999). At the same time, due to the enhancement of the immune systems, it can help people reduce the possibility of being infected by other diseases.

Between 80 per cent - 85 per cent of all medicinal mushroom products are derived from the fruiting bodies, which have been either commercially farmed or collected from the wild, *e.g.*, Lentinan, a high-molecular weight (1-3)-β-D-glucan, from *Lentinula edodes* and various products from *Ganoderma lucidum.* Only about 15 per cent of all products are based on extracts from mycelia. The notable examples include PSK-trade name (Krestin) of a polysaccharide peptide, and PSP (polysaccharide-bound peptide) extracted from *Coriolus versicolor.* A smaller percent of mushroom products are obtained from culture filtrates, *e.g.*, schizophllan, a high-molecular weight (1-3), (1-6)-β-D glucan, prepared from *Schizophyllum commune* Fr. and PSPC (a protein-bound polysaccharide complex) from *Tricholoma lobayense* Hein. However, due to increased quality control and year round production, mycelial products are the wave of the future.

Table 2: Comparison of Mushroom with some Common Vegetables

Name	*Calories*	*Moisture*	*Fat*	*Carbohydrate (per cent)*
Mushroom	16	91.1	0.3	4.4
Beet root	42	87.6	0.1	9.6
Brinjal	24	92.7	0.2	5.5
Cabbage	24	92.4	0.2	5.3
Cauliflower	25	91.7	0.2	4.9
Celery	18	93.7	0.2	3.7
Green beans	35	88.9	0.2	7.7
Green peas	98	74.3	0.4	17.7
Lima beans	128	66.5	0.8	23.5
Potato	83	73.8	0.1	19.1

The market value of medicinal mushrooms and their derivative dietary supplements worldwide was about US$1.2 billion in 1991, and about US$3.6 billion in 1994 (Chang 1996). In 1999, it was estimated to be US$6.0 billion (Wasser *et al.*, 2000). The market value of *Ganoderma-based* nutriceuticals alone in 1995 was estimated at US$1.6 billion (Chang and Buswell 1999). The corresponding monetary values were also generated by another famous mushroom, *Lentinula edodes.* Ninety nine percent of all sales of medicinal mushrooms and their derivatives occurred in Asia and Europe with less than 0.1 percent in North America. The 1999 US market for dietary supplements based mainly on mushrooms was estimated to be US$35 million. However, in recent years, the North American demand is increasing between 20 per cent -40 per cent annually, depending upon species.

Table 3: Medicinal Properties of some Mushrooms

Mushroom	*Medicinal Property*
Agaricus bisporus (Button)	Stimulating digestion, curing hypertension
Auricularia polytricha (Black ear)	Strengthening health, helping blood circulation
Boletus edulis	Causing muscles and joints to relax
Coriolus versicolar	Curing chronic diseases
Ganoderma lucidum	Rejuvenating effect, Anti hepatitis B Anti HIV, Anti tumour
Griflola forndosa (Maitake)	Strengthening health and resisting disease
Hericium erinaceus	Beneficial to heart, kidneys, lungs, liver, spleen
Lentinus edodes (Shiitake)	Strengthening health and resisting diseases
Pleurotus ostreatus (Dhingari)	Causing muscles and joints to relax
Tremella fuciformis (Silver ear)	Strengthening health and resist diseases
Tricholoma mongolicum	Beneficial to stomach and intestines

Chapter 5
Mushroom Recipes

5.1 Soups

- ☆ Mushroom soup
- ☆ Mushroom tomato soup
- ☆ Cream of mushroom soup
- ☆ Mushroom mix vegetable soup
- ☆ Cream of oyster mushroom
- ☆ Mushroom onion soup
- ☆ Mushroom egg soup

5.2 Salads

- ☆ Stuffed tomatoes with mushroom
- ☆ Fresh pickled mushrooms
- ☆ Mushroom egg salad
- ☆ Sprouts with Mushrooms

- ☆ Mushrooms cabbage salad
- ☆ oriental mushroom salad

5.3 Sauces

- ☆ Mushroom tomato sauce
- ☆ Mushroom white sauce
- ☆ Mushroom brown sauce
- ☆ Mushroom onion sauce

5.4 Snacks

- ☆ Mushroom cabbage rolls
- ☆ Mushroom pakora
- ☆ Mushroom shami kabab
- ☆ Mushroom sandwich
- ☆ Mushroom kabab
- ☆ Mushroom omelette
- ☆ Mushroom bread pakora

5.5 Main course

- ☆ Mushroom mixed vegetable
- ☆ Mushroom with baby corn in gravy
- ☆ Baked mushroom masala
- ☆ Kadhai mushroom
- ☆ Mushroom dum biryani
- ☆ Mushroom mutter
- ☆ Mushroom Curry
- ☆ Mushroom fry
- ☆ Garlic mushroom
- ☆ Tawa mushroom

5.6 Chinese Recipes

- ☆ Mushroom noodles
- ☆ Mushroom with vegetable Manchurian
- ☆ Mushroom Manchurian
- ☆ Mushroom fried rice
- ☆ Mushroom and peas pasta

Chapter 6
Types of Mushroom

6.1 Cultivated Types of Mushrooms

1. Button mushrooms: *Agaricus bisporus*
2. White button mushrooms: *Agaricus bitorquis*
3. Oyster mushrooms: *Pleurotus* spp.
4. Paddy straw mushrooms: *Volvariella* spp.
5. Milky mushrooms: *Calocybe indica*
6. Black ear mushrooms: *Auricularia* spp.
7. Shiitake: *Letinula edodes*

6.2 Wild Types of Mushrooms

1. *Morchella* spp (Guchhi)
2. *Boletus* spp
3. *Cantharellus* spp
4. Tuber mushrooms

4. *Volvareilla volvacea:* Paddy Straw Mushroom

- ☆ It is also known as Chinese mushroom
- ☆ It requires a temperature of 33-37°C and 85-95 per cent RH for fruiting

- ☆ Grow well on paddy straw
- ☆ Major constrain is low yield and poor shelf life

5. *Calocybe indica*: Milky Mushroom

- ☆ It was first reported from India in1974

- It grows in nature on humus soil under the road side trees or in the forest
- It grows well at higher temperature of 30-40° C
- It grows both on wheat straw and paddy straw
- Artificial technique has been standardize and adopter by farmers in Tamil Nadu and Karnataka.

6. *Auricularia auricularia*: Black Ear Mushroom

- It is the oldest mushroom grown by Chinese
- Fruiting of this mushroom occur s best at 27° C

- It can be grown on both wheat and paddy straw
- Average yield of this mushroom is 800-1200 g fresh mushroom per kg of dry straw.
- 15-20 tonnes of dry mushrooms exported annually from India

7. Shiitake: *Lentinus edodes*

- This is a medicinal mushroom being cultivated on large scale in China and Japan
- It requires low temperature of 10-20 °C

☆ NRCM has achieved success in growing this mushroom on artificial wood log made of saw dust is in its final stage

8. *Morchella deliciosa*: Wild Edible Mushrooms

☆ This is the most prized mushroom presently collected from high hills of J&K, H.P. and Uttaranchal

- ☆ Most of it is exported to European countries
- ☆ Extensive research is going on to domesticate this mushroom

9. *Boletus* spp: Wild Edible Mushroom

- ☆ It has a central stipe with fleshy cap whose lower surface is composed of vast numbers of pores.

- ☆ Some of these can reach a diameter of 2 ft. and weigh 3 Kg with impressive colour and texture
- ☆ Most of the *Boletus* spp. are edible

10. *Cantharellus* spp: Wild Edible Mushrooms

- ☆ A very popular edible fungus which form funnel shaped fruit body
- ☆ It lacks true gills and is found in damp, mossy woodlands, either in pine or deciduous

11. Tuber Mushrooms: Wild Edible Mushroom

- Also known as truffles
- These are found in deciduous woods
- They have a distinctive smell and have good commercial value
- The size may vary from the size of walnut to that of potato
- Most of them are edible

12. Poisonous Mushroom: *Amanita muscarria*

- ☆ They are known as toadstool
- ☆ No general guidelines for distinguishing edible and poisonous species of mushrooms
- ☆ Do not eat a mushroom because it is beautiful and has a pleasant smell

- ☆ Newer eat uncooked mushrooms unless you are absolutely sure that it is edible
- ☆ Avoid gathering and eating wild button mushrooms, at button stage species cannot be distinguish easily.
- ☆ Avoid mushrooms oozing a white milky juice when cut.
- ☆ When eating an edible species for first time eat only in small amounts.
- ☆ Do not eat over-mature, insect infested, or wilting wild mushrooms
- ☆ Mushrooms partially eaten by animals or insect may not be safe for human consumption

Chapter 7

Mushroom Production Technology

- ☆ Preparation/Sterilization of substrate
- ☆ Filling of substrate in container
- ☆ Spawning/Spawn Preparation
- ☆ Removal of bags
- ☆ Harvesting and Marketing.
- ☆ Yield and Economics of Mushroom cultivation
- ☆ Storage of mushroom

7.1 Mushroom Spawn Production

In nature, mushroom spores help in survival of a species from one generation to the next. These spores are extremely small microscopic propagules and therefore difficult to handle as seed. Spores need time and specific condition to germinate and during this time competitor fungi might germinate and grow faster to exhaust the available substrate. Therefore, a pure culture of the desired mushroom mycelium is first raised on a convenient artificial culture medium and then added to the substrate to give it an advantage. The spawn comprise mycelium of the mushroom and a supporting medium, which provides nutrition to the fungus during its growth. Spawn is used as inoculums or 'seed' for the substrate in mushroom cultivation. Right kind and quality of spawn is very important for cultivation.

The isolation, purification and maintenance of mushroom cultures require technical expertise and aseptic high-tech laboratory facilities. Therefore, small mushroom growers cannot maintain their pure culture and or spawn. They have to rely entirely on commercial spawn producers, reliable governmental and non-governmental organization that play a veal role in supplying reliable spawn of a desired mushroom strain or variety.

a) Preparation of Culture Media

Besides infrastructure, raw material and equipment, healthy vegetative mycelia culture is required to produce quality spawn. The vegetative mycelium is raised on to a convenient culture media. A variety of culture media can be used to grow vegetative mycelium before it is inoculated on a suitable substrate to grow mushrooms. These culture media are also used as substrate for isolation, multiplication, maintenance and preservation of mushroom cultures. For the convenience of the practical users a few compositions of different media are given below, which can be used for making slants or Petri dishes.

Malt Extract Agar (MEA)

Malt extract	: 25 g
Agar-agar powder	: 20 g
Water	: 1 liter

Water is boiled, mixed with malt extract and the ingredients are dissolved using clean glass rod by stirring water to avoid limp formation. The pH of the medium is adjusted between 6.5 to 7.0 before autoclaving.

Wheat Extract Agar (WEA)

Wheat grains : 32 g

Agar-agar powder : 20 g

Water : 1 liter

pH : 7.0

Wheat grains are boiled in water for 1-1½ hours and are filtered through the muslin/cheese cloth and discarded. The volume of grain extract is raised to 1 liter with water and then agar-agar powder is added by stirring continuously and PH is adjusted.

Potato Dextrose Agar

Peeled and diced potato: 200 g

Dextrose : 20 g

Agar-agar powder : 20 g

pH : 7.0

Potatoes are peeled, washed, cut into small pieces (3-4) and boiled in water for 20-25 minutes. The potato extract is filtered through a cheese cloth and potato cubes discarded. The volume of this extract is raised to one 1 liter. Dextrose and agar-agar powder is then mixed by stirring and pH is adjusted. Green potatoes should not be used as they contain anti-fungal alkaloids, which may be harmful to mushroom mycelium.

Compost Extract Agar

Pasteurized and conditioned compost	: 150 g (Fresh weight)
Agar-agar powder	: 20 g
Water	: 1 liter
pH	: 7.0

Pasteurized compost is boiled in 1½-2 liter of water until the volume of water is reduced to half. It is filtered through cheese cloth and the volume of the compost extract is raised to one liter. Agar-agar powder is mixed by stirring and pH adjusted. Streptomycin sulfate @ 50mg/liter may be added after autoclaving to eliminate bacterial contamination.

Oat Meal Agar

Oat meal flakes	: 30 g
Agar-agar powder	: 20 g
Water	: 1 litre
pH	: 7.0

Oat meal flakes are boiled in sufficient water for 2 hours. The volume of the supernatant is raised to one liter. Agar-agar powder is mixed by stirring and pH is adjusted.

Rice Bran Decoction Medium

Rice bran	: 20 g
Agar-agar	: 20 g
Water	: 1 litre
pH	: 7.0

Rice bran is boiled in water for 15-20 min, filtered through cheese cloth and then agar-agar powder is mixed by stirring. pH is adjusted.

If the pH of the medium is below 6.5, $N/_{10}$ NaOH is added drop by drop to raise it to 7.0 whereas; if the PH is above 7.0 it can be brought down by adding $N/_{10}$ HCl. The pH is to be adjusted to 7.0 before sterilization. It has been observed that the pH declines approximately 0.5 upon autoclaving. After autoclaving the pH comes down to 6.5 levels, which is most suitable for mushroom mycelium to grow. If the culture are to be raised in petri plates then the medium is first sterilized in 500ml conical flasks,15-20 ml of lukewarm sterilized culture medium is poured in each pre-sterilized petri plates. If the cultures are to be multiplied in test tubes then culture is poured in large number of culture tubes @ 15 ml/tube. The culture tubes or conical flasks are then plugged with non-absorbent cotton and autoclaved under moist steam at 15 pounds per square inch (p.s.i.) pressure for 25-30 minutes. The culture medium is allowed to solidify in culture tubes or in petri plates for a few hours before they are used for inoculation or sub-culturing. The petri plates are sterilized in an oven at 180°C for 2½ hours and allowed to cool down to room temperature before they are used for planting media.

b) Isolation from Fruit Body

We need sterilized culture medium both in petri plates and culture tubes, scalpel, and inoculation needle, wide mouth test tubes, spirit lamp, scissors, forceps, matchbox, and ethanol in an inoculation room. Ultra violet (UV) lights should be put on for about half an hour before working in the laminar flow. All material except the living cultures and fruit body should be exposed to UV light to eliminate contamination present on the surface of culture tubes, forceps etc. Before starting the inoculation, the surface area of inoculation table should be swabbed with cotton dipped in rectified spirit. After putting off UV lights hands should be washed with rectified spirit and air dried before lighting spirit lamp to avoid accidents.

Pure cultures of fleshy fungi/mushrooms can be prepared either by tissue culture or multi-spore culture.

Tissue Culture

In case of *Agaricus* species, a button–stage young fruit body should be freshly harvested and brought to the laboratory for tissue culture. The fruit-body should be cleaned and adhered casing soil may be removed with the help of cotton swab. The fruit-body is held with the help of scalpel/inoculation needle and dipped in 0.1 per cent mercuric chloride ($HgCl_2$) or 70 per cent ethanol and then heating it on flame. After cooling scalpel, the fruit-body is split longitudinally and tissue bits of approximately 3-5mm are cut from pileus-stipe junction (collar region). The bits are aseptically transferred onto pre-sterilized petri plate containing convenient culture medium with the help of a sterilized forceps. These petri plates are wrapped with parafilm and incubated in a BOD incubator at 25°C±1°C for 6-7 days. The actively growing mycelium from edges along with culture medium of about 5mm size should be transferred to a number of test tubes. These test tubes are then incubated in a BOD incubator at 25°C±1°C for 2-3 weeks. This pure culture can then be used as inoculums for spawn preparation. In case of *Pleurotus* spp., the tissue culture is raised from the junction of pileus and stipe portion. Rest of the procedure is same. In case of tropical mushrooms like *Volariela* and *Calocybe* species the petri plates and test tubes containing bits of fruit-body are incubated at 30-32°C instead of 25°C.

Multi spore Culture

Multi spore culture is also used for raising pure culture. It is raised form a mass of germinated basidiospores. From a spore print, a loop full of spore mass is picked using sterilized inoculation needle, suspended in 10ml sterile distilled water in 18 x 150 mm test-tube, and mixed thoroughly. One ml of this suspension is remixed in nine ml sterile water. Again, one ml of diluted spore suspension is spread with the help of glass rod on to a petri plate containing 20 ml of a convenient culture medium. These petri plates are incubated in BOD incubator at 25ºC+1ºC for one week. The fast growing multi-spore colonies are picked up with piece of agar-agar and transferred to freshly prepare convenient culture medium in test tubes under aseptic conditions. These test tubes are incubated for 2-3 weeks in a BOD incubator. Multi spore culture is then ready for inoculation in the wheat grain substrate for spawn preparation.

c) Sub-Culturing

As a result of isolation and purification, a pure vegetative mycelia culture is established. These cultures are maintained and multiplied for their use in inoculation of grain substrate for large-scale production of spawn and use in future for variety of purposes. Sub-culturing is done by transferring a small piece of growing pure culture along with culture medium on a suitable medium. The compositions of most commonly used media have already been described. In order to maintain vigour of the mycelium the culture media may be changed in subsequent sub-culturing.

Spawn Production

a) Substrate

Cereal Grains

Mushroom spawn can be prepared on any kind of cereal grains like wheat, maize, bajra, jowar or rye etc. The large grains carry a greater reserve of food material per grain to sustain the inoculums of mushroom mycelium until it is established and feeding on the compost, so they may be more effective in poor compost or adverse conditions. Whereas, the small grains provide more points of inoculums per gram of spawn, so if all the grains of both types grow equally well, the small ones will penetrate the compost sooner.

Other Substrates

Although cereal grains are suitable for making spawn of any mushroom verity but due to prohibitive cost a variety of agricultural wastes like corn cobs, wooden sticks, rice straw, sawdust and used tea leaves etc. have also been used.

Sawdust of red wood tree species is preferred for making spawn. Wheat bran ranging from 10 to 20 per cent is mixed with sawdust to increase moisture percentage. Used tea leaves are washed and dried before using them for spawn preparation. Ipil ipil leaves are mixed with saw dust in the ratio 3:1 and used as substrate.

b) Substrate Preparation

Spawn substrate should have following desirable characteristics:

- ☆ It should not contain any inhibitory compounds to desirable mushroom species.

- ☆ Large surface area of substrate should be available for fungal colonization.
- ☆ It should provide essential nutrients required by mushroom mycelium to grow.
- ☆ Cereal grains should be free from diseases.
- ☆ Cereal grains should not be broken, old or damaged by insects.

Spawn preparation using cereal grains as substrate is described in details. The cereal grains are thoroughly washed in sufficient water 3-4 times to remove soil debris, straw particles and undesirable seeds and grasses etc. washed grains are then soaked in sufficient water for 20-30 minutes to save energy. These grains are taken in a wide mouth container and boiled in sufficient water for 20-25 minutes. Normally, for soaking and boiling 20 kg of wheat grains, 35 litre of water is required. After boiling, grains should absorb 55-60 per cent moisture. Excess water from the boiled grains is removed by spreading on sieve made of fine wire mesh or muslin cloth. The grains are allowed to leave as such for a few hours so that the water on the surface is evaporated. Now the grains are mixed with Gypsum (Calcium sulphate. $CaSO_4$) and chalks powder (Calcium carbonate $CaCO_3$) so that the PH of the grains is around 7 to 7.5 and they do not form lumps. Different people have given different ratios for mixing gypsum and calcium carbonate. The best results have been obtained by using 200 g gypsum and 50 g calcium carbonate for 10 kg grains. These calculations are on dry weight basis of grains used for spawn preparation. First gypsum and calcium carbonate are mixed separately and then the mixture of both is thoroughly mixed with the grains. This mixing should be done on a smooth surface after wearing gloves to avoid contamination.

Mother/Master Spawns Preparation

About 300 g prepared substrate (boiled cereal grains coated with gypsum and calcium carbonate) is filled in glucose/milk bottles up to 2/3 volume and plugged with non-absorbent cotton. These bottles are then autoclaved at 22 lb p.s.i. pressure at 126°C for 2 hours. These autoclaved bottles are left in the room for 24 hours so that they are cooled to ambient temperature and the excess moisture accumulated inside the bottles walls is evaporated. These autoclaved bottles containing sterilized grains are then exposed to UV light inside the inoculation chamber for about 30 minutes. A piece of growing mycelium is aseptically transferred to these bottles and inoculated bottles are incubated at 25°C. Inoculated bottles are gently shaken in 5th and 10th day after inoculation. Fully colonized mother spawn bottles can be used for inoculating commercial spawn bags after two to three weeks. Inoculated bottles are inoculated at 22-25°C for *Agaricus, Pluerotus, Lentinula* and *Auricularia* species of *Volvariella* and *Calocybe.*

Commercial Spawn Preparation

Commercial spawn can be prepared in heat resistant polypropylene bags. Normally for half and one kg spawn, the bags should be of 35x17.5 cm and 40x20cm size, respectively. Polypropylene bags should have double sealing at the bottom and after filling the grains they are plugged with the help of a polypropylene neck and non-sorbent cotton. The polypropylene bags of 150 gauges should be used. The bags are then sterilized at 222 lb p.s.i. pressure for 1.5 to 2 hours in an autoclave. Autoclaved bags are shacked well before inoculation so that the water droplets accumulated inside the bags are reabsorbed by the grains. The sterilized bags are kept in the laminar flow under UV light for about 30 minutes. Ten to fifteen grams of grains from master spawn bottles is inoculated per bags under aseptic conditions. One bottle of master spawn is sufficient for inoculating 25 to 30 commercial spawn bags of half kg capacity. Inoculated bags are again shacked so that the inoculum is well mixed with other grains. However, to reduce the time period for spawn preparation, the quantity of inoculums may be increased. Then the bags are kept in incubation room or in BOD incubator for mycelium spread at 25°C. During incubation the bags are regularly examined for mould infestation, contaminated bags should be immediately removed and autoclaved before discarding the bags or avoid build up of contamination in the vicinity. Normally it takes 20-25 days for complete spread of mycelium on the grains for *Agaricus* species.

Spawn Storage and Transport

If possible, freshly prepared spawn should be used while the mycelium is in the state of active growth. The spawn bags after completion of log growth phase can be maintained up to 3-4 months at 4°C. The commercial spawn should be systematically packed in ventilated bags or cardboard boxes. The spawn should be carried in refrigerated vans to large distances. Alternatively, it can also be transported to long distances during night by public transport or private vehicles so that the temperature of planter spawn do not rise beyond 30-32°C. The spawn of *Volvariella* and *Calocybe* should not be stored at refrigerated temperatures as these mushrooms are sensitive to low temperatures. Old spawn bags of *Auricularia* may also ooze brownish liquid at cooler temperatures that may result in crop losses.

Preparations of Mother/Master Spawn Culture

Step-1: Healthy and cleaned cereal grains

↓

Step-2: Boil grains in water (20-25min.)

↓

Step-3: Remove excess water through sieve

↓

Step-4: Dry grains in shade (4-6 hrs.)

↓

Step-5: Mix $CaCO_3$ (0.5 per cent) and $CaSO_4$ (2 per cent)

↓

Step-6: Fill 300g grains in each glucose/milk bottle

↓

Step-7: plug cotton and autoclave at 22p.s.i for 2 hr.

↓

Step-8: Inoculate mycelium of desired strain using laminar flow

↓

Step-9: Inoculate in BOD at 25°C for 20-25 days (Tropical mushrooms at 30-32°C)

↓

Step-10: Mother/Master Spawn are ready.

Preparation of Commercial spawn

Step-1: Use polypropylene bags instead of bottles for grain filling

↓

Step-2: Up to autoclaving step 1 to 7 are same as of master spawn

↓

Step-8: Inoculate about 10 gm from mother spawn to polypropylene bag containing grains

↓

Step-9: Shake bags after 7-8 days

↓

Step-10: Incubate at 25° C in incubation room or BOD (Tropical mushrooms at 30-32°C)

↓

Step-11: Commercial spawn is ready in 20-25 days.

Precautions in Spawn Preparation

1. Strict hygiene should be maintained throughout right from raising of pure culture to storage and transport of spawn
2. Excess of visitors to inoculation room should be denied.
3. Inoculation room should be disinfected regularly by exposing it to formalin or by installing ozone generator to eliminate chances of contamination.
4. During incubation of spawn, bottles and bags should be inspected frequently to remove contaminated bags.

5. To avoid spread of contamination in the vicinity, the contaminated bags, must be autoclaved and buried in the soil.
6. Before spawning into the substrate the stored spawn under refrigerated conditions should be brought to ambient temperatures.
7. It is necessary for each growing cycle to start with a fresh spawn.
8. Regular floor cleaning with surface disinfectants like dettol, Lysol etc. should be done twice a week.
9. Cultures should be wrapped with aluminum foil to prevent contamination of cultures during storage under refrigerated conditions.
10. Cultures and spawn should never be stored ant sub-zero temperatures in deep freezers. However, cultures can be preserved at ultra low temperatures using cryo-protectants following proper protocols already described separately.
11. Over autoclaving of culture medium beyond the prescribed pressure and time should be avoided as it may cause charamalization of sugars in the medium.

7.2 Cultivation of *Agaricus bisporus*

Agaricus bisporus is variously known as the *white mushroom, button mushroom, champignon,* or simply the *common cultivated mushroom.* In Western countries this mushroom has developed over the past 60 years from beginning as a risky venture to a largely predictable and controllable industrial process, particularly in Great Britain and the Netherlands. In no small measure this remarkable achievement in modern mushroom industrial development may be attributed to contributions resulting from the vigorous research activities conducted at mushroom research laboratories, centers and stations.

The basic principles and practical methods of cultivation of this mushroom have been very well established through repetitive practical experiences during the last 60 years. Generally, composting refers to the piling up of substrates for a certain period of time and the changes due to the activities of various micro-organisms, which result in the composted substrate being chemically and physically different from the starting material. This is sometimes referred to as a solid state fermentation. Two types of composting are commonly described. One type involves the decomposition of heaps of organic wastes and the subsequent application of the residue to the soil. The aim of this type of composting is to reduce, in a sanitary manner, both the volume and the C:N ratio of the organic waste so that is it suitable for manuring the soil to improve the growth of plant crops. When given directly to the soil without composting, organic waste with a high C:N ratio (such as straw) can give rise to a temporary nitrogen deficiency which will then result in a reduction in yield of the plant crop.

The second type of composting is also a process of microbial fermentation, but in this case the substrate is used for the cultivation of edible mushrooms. Through composting, a mixture of rich organic materials is converted into a stable medium which is selective for the growth of a particular mushroom but is not suitable, or is less favourable, for the growth of competing micro-organisms. The competitors exist in uncomposted materials, and often in partially composted materials, but they are far less active in well-composed mushroom substrates. Actually, this type of composting is derived mainly from the *Agaricus* mushroom-growing industry, in which a composting technique which renders wheat straw with horse manure specific for the growth of the *Agaricus* mushroom has been developed. It should be noted that the treatment of substrates for growing other mushrooms can be regarded as "composting", but the procedures followed in composting and the nature of the product can be quite different. This is because the starting materials and the lengths of time accompanying the various changes in the substrates vary from mushroom to mushroom.

As stated above, the role of composting is the production of a selective substrate that will preferentially support the growth of the mycelium of the mushroom. The basis of this selectivity, however, cannot be attributed to one factor or one aspect of the entire system. The physical, chemical and biological aspects of composting are fundamentally interrelated, but can be artificially separated for the convenience of investigation and discussion.

Mushroom growers like to practice the liberal use of the sense of sight, smell and touch to evaluate the progression of the composting process and the quality of the final product. The gross characteristics of compost, usually referred to as "structure", result from a number of complex physical, chemical and microbial processes.

The overall goal of composting is to produce selective nutrient media for the growth of the mushroom. These selected nutrient-rich substrates should support a high yield of good quality mushrooms. The general aspects of the achievement of composting are summarised as involving:

1. Straw softening and other structural changes;
2. Modification of plant materials so that nutrients are made available to mushroom growth and development;
3. Building up of an appropriate biomass and a variety of microbial products(some of these can serve as nutrient sources for the mushroom);
4. Establishment of selectivity, *i.e.* the compost promotes the growth of the mushroom over competitor organisms;
5. Modification of compost structure so that it holds more water; and
6. Building up of compost moisture content to serve as a water reservoir for the mushroom crop.

Composting is prepared in accordance with well-documented commercial procedures (van Griensvan, 1988, Chang and Hayes, 1978, and Kaul and Dhar, 2007). In Phase I of the process (outdoor composting), locally available raw materials are arranged into piles which are periodically turned and watered. The initial breakdown of the raw ingredients by micro-organisms takes place in Phase I. This phase is usually complete within 9-12 days, when the materials have become pliable, dark brown in colour and capable of holding water. There is normally a strong smell of ammonia. Phase II (indoor fermentation) is pasteurization, when undesirable organisms are removed from the compost. This is carried out in a steaming room where the air temperature is held at 60°C for at least 4 hours. The temperature is then lowered to 50° C for 8 to 72 hours depending upon the nature of the compost. CO_2 is maintained at 1.5 to 2 per cent and the ammonia level drops below 10 PPM. Following Phase II composting, the substrate is cooled to 30° C for *A. bitorquis* and to 25° C for *A. bisporus* for spawning. Production of Phase III or Phase IV composts for growing *Agaricus* mushrooms has been an advanced technological development in recent years in Western countries. The production of Phase III compost is Phase II compost spawn run in a bulk tunnel, and ready for casing when delivered to the grower. If the Phase III compost is then cased and spawn developed into casing layer before dispatching to the growing unit or delivering to growers, it is named as Phase IV compost. The successes of bulk Phase III and Phase IV depend a lot on the quality of Phase I and Phase II processes. Phase II on shelves produce an average of 4.1 crops per year. Since 1999, growers using Phase III production enjoyed an average of 7.1 crops per year. In recent years, Phase IV can generate 10-12 crops per year (Dewhurst 2002, Lemmers 2003).

7.3 Cultivation of *Pleurotus sajor-caju*

Pleurotus sajor-caju (grey oyster mushroom) is comparable to the high temperature species in the group of *Pleurotus* (oyster) mushrooms, with high temperatures required for fructification. This mushroom has a promising prospect in the tropical/subtropical areas. Its cultivation is easy with relatively less complicated procedures (Chang and Miles, 2004, Kaul and Dhar, 2007).

1. *Biological nature*: The temperature for growth of mycelium is 10-35°C. The optimum growing temperature of the mycelium is 23-28°C. The optimum developmental temperature of the fruiting body is 18-24°C. The optimum pH of the

substrate used in making the mushroom bag/bed is 6.8-8.0. The C:N ratio in the substrate is in the range of 30-60: 1. A large circulation of air and reasonable light are required for the development of the fruiting bodies.

2. *Examples of spawn substrates*: (i) Wheat grain + 1.5 per cent gypsum or lime. (ii) Cotton seed hull 88 per cent, wheat bran 10 per cent, sugar 1 per cent and gypsum 1 per cent. (iii) Sawdust 78 per cent, wheat bran 20 per cent, sugar 1 per cent and gypsum 1 per cent. (vi) Sawdust 58 per cent, spent coffee grounds/spent tea leaves 20 per cent, water hyacinth/cereal straw 20 per cent, sugar 1 per cent and gypsum 1 per cent.
3. *Examples of cultivation substrates*: (i) Cotton seed hull 95 per cent, gypsum 2 per cent, lime 1 per cent and Calcium superphosphate 2 per cent. (ii) Rice straw 80 per cent, cotton waste 18 per cent, gypsum 1 per cent and lime 1 per cent. (iii) Water hyacinth 80 per cent, cereal straw 17 per cent, gypsum 2 per cent and lime 1 per cent.

For demonstration purposes, this mushroom can be nurtured to grow into a tree-like shape (Chang and Li, 1982). The cultivation method, which has been tested to be successful, is as follows: Cotton waste or rice straw mixed with water hyacinth is used as the substrate. Tear large pieces of cotton waste into small parts or cut the straw and water hyacinth into small segments. Add 2 per cent (w/w) lime and mix with sufficient water to get moisture content of about 60-65 per cent. Pile the materials up, cover with plastic sheets and leave to stand overnight. Load the substrate into small baskets or on shelves for pasteurisation or cook the substrate with boiled water for 15 minutes. After cooling to approximately 25°C, mix around 2 per cent (w/w) spawn thoroughly with the substrate and pack into columns of 60 cm long tubes which have hard plastic (PVC) tubing of 100 cm (4 cm in diameter) as central support, and with plastic sheets as outside wrapping.

Incubate these columns at around 24-28°C, preferably in the dark. When the mycelium of the mushroom has ramified the entire column of substrate after three to four weeks, remove the plastic wrapping and switch on white light. Watering occasionally is needed to keep the surface from drying. In around three to four days white primordia start to appear over the whole surface. After another two to three days, the *Pleurotus* mushrooms are ready for harvesting. During the cropping period watering is very important if many flushes are required.

7.4 Cultivation of *Volvariella volvacea*

The edible straw mushroom, *Volvariella volvacea* is a fungus of the tropics and subtropics and has been traditionally cultivated in rice straw for many years in China and in South East Asian countries. In 1971, cotton wastes were first introduced as heating material for growing the straw mushroom (Yau and Chang 1972), and in 1973, cotton wastes had completely replaced the traditional paddy straw to grow the mushroom (Chang 1974). This was a turning point in the history of straw mushroom cultivation, because the cotton-waste compost through pasteurisation process brought the cultivation of the mushroom into an industrial scale first in Hong Kong and then in Taiwan, Thailand, and China. Several techniques are adopted for

Oyster Mushroom Cultivation Flow Chart

Raw material

↓

(Paddy straw or vegetable plant residue)
Chopping (3.5 cm)

↓

Soak in water (8-24 hours)

↓

Polythene bag method	Gunny bag method
Fill in polythene bags	Fill in gunny bags
↓	↓
Pasteurization at 15 lbs/sq. inch for 60 minutes	Dip in hot water (80-85° c) for 10 minutes
↓	↓
On cooling do surface spawning (2 per cent) inside sterile chamber @ 2 per cent	On cooling fill bags in 3-4 layers each of 10cms with spawning

↓

Spawn running (20-25 days)

↓

Pinning (3-5 days)

↓

Cropping 1st flush (3-4 days duration)

↓

Subsequent flushes at 10 days intervals

↓

Pickling (20-30 days)

Flow Chart of *Volvariella volvacea*

Substrate (Paddy straw)

↓

Bundle 1 kg each 12-15 cm diameter

↓

Soaking in water 8-16 hours

↓

Pasteurization in hot water 80-85°C for 10-15 minutes

↓

Bed preparation and spawning 34°c

↓

Cover the bed with polythene sheet 150 gauge

↓

Spawn running 32°C -34°C (8-10 days)

↓

Pinning 29°C-32°C (3-4 days)

↓

Harvesting 29°C-32°C (9-10 days)

Milky Mushrooms Cultivation Flow Chart

Sterilize substrate (wheat or paddy straw) by Bavistin/Hot water or steam

↓

Add organic supplement maize meal, wheat bran, rice husk or lucern hay @ 5 5 and 7.5 per cent at spawning and casing, respectively

↓

Spawn @ 2-3 per cent

↓

Spawn run completed in18-20 days at 20-37° C

↓

Casing with 2 years old cow dung patties or F.Y.M.+ spent compost (Sterilized) to a thickness 3-4 cm for 10-12 days

↓

Fruiting primodia start developing in 5-6 days (Temp. required 25-30 °c for fruiting)

Black Ear Mushrooms Cultivation Flow Chart: *Auricularia* spp

Wheat straw + 4 per cent rice bran supplementation

↓

Sterilization of substrate 15 lbs for 1-2 hours

↓

Fill the bag and spawn @ 2 per cent

↓

Spawn running – 2 weeks

↓

Fruiting within 20-25 days

↓

Total developmental period 40-60 days

↓

Good biological efficiency 80-130 per cent

the cultivation of the mushroom, which thrives in the temperature range of 28-36° C and a relative humidity of 75-85 per cent. Detailed descriptions of the various methods

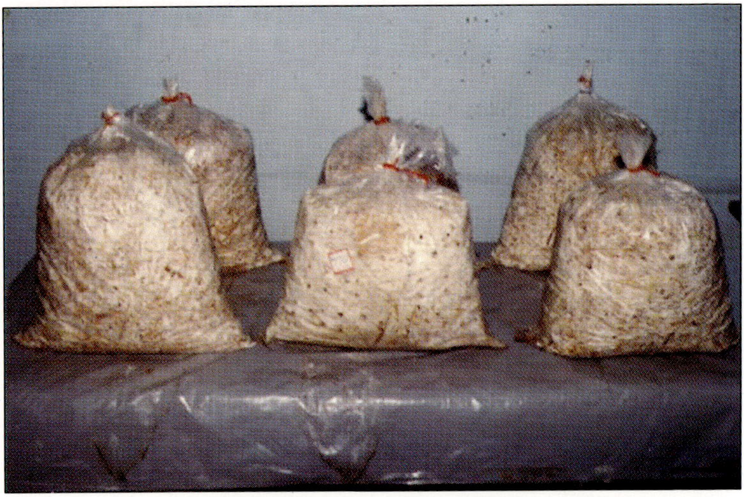

Bagging and Spawning

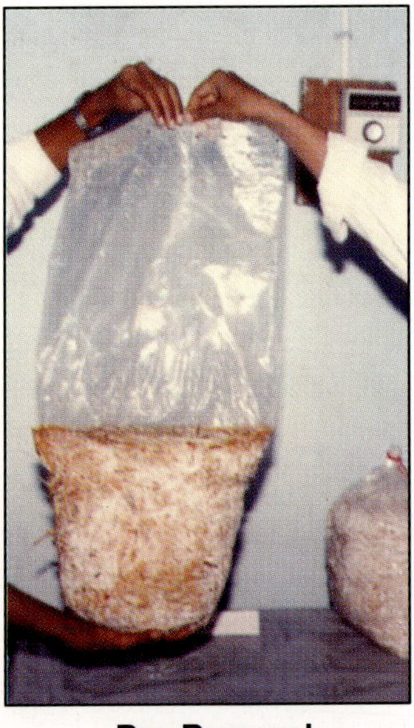

Bag Removal

Spawn Run

Pin Head

Mature

Harvesting

are given by Chang and Quimio (1982), Chang and Miles (2004), Kaul and Dhar (2007) and Quimio, *et al*. (1990). Choice of technology usually depends on personal preference, and on the availability of substrates and the amount of resources available. While the more sophisticated indoor technology is recommended for an industrial-scale production of the mushroom, most of the other technologies are low-cost and appropriate for rural area development, especially when production is established at the community level.

Chapter 8
Insect Pests of Mushroom

Like the field crops, mushrooms are attacked by several pests, sciarids flies, phorid flies, cecids, springtails, mites and nematodes are important pests of cultivated mushrooms throughout the world. These pests damage the crop right from spawning to harvesting of the crop. Mushrooms being the indoor crop provide a very suitable habitat for the insect pests due to which pest remains protected from the vagaries of the weather. Mushrooms flies, besides spreading various mushroom diseases, take heavy toll of the crop every year. Therefore, measures should be taken to prevent the entry of insect pests, mite and nematode into the cropping rooms. However, chemical method of pest control should be taken as last step rather than the first. The important insects, mites and nematodes pests of buttons mushroom are discussed below.

8.1 Sciarids Fly

Sciarids are major pest of mushroom throughout the world. These are delicate flies of small or medium size, often bearing resemblance to gnats or midges and exceedingly numerous in individuals and species. Upwards of 2000 species are known and the geographical range of the family is very wide.

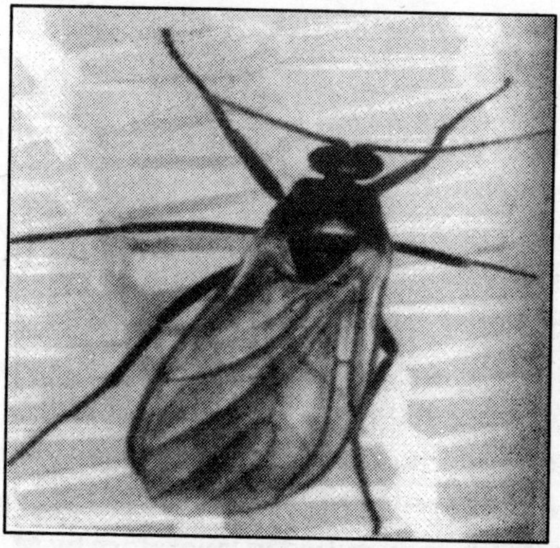

Sciarids can cause up to 50 per cent loss in yield. Damaging stage of the pest is larvae. Larvae feed on compost, mushroom mycelium and mushrooms. Through the consumption of compost by the larvae, pH of the substrate changes which slows down the growth of mushroom mycelium. As the infestation by the larvae is often in groups, bare patches without mushrooms can be seen on the beds. After the sporophore formation, larvae enter the mushrooms and start feeding and tunnel within a stipe. Eventually they reach the pileus and feed vigorously. When larval attack occurs at pin head stage, further development of pinheads completely stops and pins eventually die. *Agaricus bisporus* is more susceptible than *A. bitrquis.* Sciarids are known to spread many mushroom diseases. They act as vector of *Veticillium fungicola* and bacteria carrying mites. Under laboratory conditions, sciarids were found to transmit 84-100 per cent. 80-100, 100 per cent, 76-100 per cent and 4-100 per cent *V. fungicola, Sepedonium chrysosprmum, Trichoderma viride, Papulosporea byssina* and *Mycogone perniciosa,* respectively.

8.2 Phorids

These are small hump backed black or light to dark brown flies measuring 1.6-2.0 mm in size. These flies are diminutive of house flies. These flies move rapidly with jerky movements. In case of button mushroom phorids can cause up to 46 per cent loss in yield. In oyster mushrooms, particularly during rainy season phorids have been observed to cause 100 per cent loss in yield. Phorids flies are also known to transmit various diseases of mushrooms. Dry bubble disease of mushroom caused by *Verticillium fungicola* is readily transmitted in laboratory by *M. malterata.*

Phorid, may cause greater economic threat as a vector of the sticky spores of *V. fungicola* than by the damage to mushroom mycelium caused by its larvae. Under laboratory conditions, *Megaselia* sp. was found to transmit 76-100 per cent of *V.fungicola,* 100 per cent each of *Trichoderma viride.* Phorids also act as vectors of mites.

8.3 Cecids

Cecids are orange coloured larvae of *Mycophila speyeri* damaging button mushrooms. Adult cecid is dark brown, tiny midge 0.7-1.5 mm. Larvae are spindle shaped, 1.5-2.8 mm white or orange in colour.

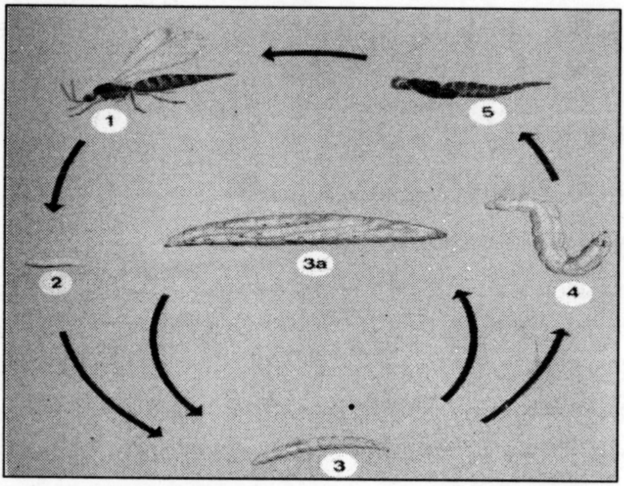

Larvae feed on mycelium and make vertical grooves in the stipe. They feed on the outside of stipe at the junction of stipe and gills. Presence of bacteria on the skin causes brown discoloured strips on the stipe and gills. Delicates gill tissues break down to produce tiny pustules of black fluid. Young larvae also feed on exuding sap. Cecids can cause up to 50 per cent spoilage of mushroom.

8.4 Springtails

Springtails cause serious damage to oyster mushroom in Rajasthan and button mushroom in Punjab. Adults are silver grey to ground colour with light violet band along including appendages is 0.7-2.25 mm. Abdomen is 4-6 segmented. An antenna is 3-6 segmented.

Springtails damage oyster, button, shiitake and milky mushrooms. The springtail feed on mycelium in compost resulting in disappearance of mycelium from spawned compost. They also attack fruiting bodies of button mushroom and cause slight pitting or browning at feeding sites. In oyster and shiitake they feed on grills resulting in destruction of gill linings. They also congregate at the base of stipe and eat mycelia strands.

Management

Physical Methods

1) Hygiene and Sanitation

Hygiene is the primary method of pest control in mushroom farming. It is the foundation upon which success of all other control techniques depends. The objectives of any hygiene programme include exclusion of pests and diseases from production cycle, elimination of pests and pathogens and destruction of pest and disease present in a crop at its termination. Such measures help to reduce the contamination level and ensure clean start for subsequent crops. Sanitation focuses on elimination or killing a pest. Routinely removing stumpage from the rooms where the crop is growing is a sound sanitary practice. Sanitary practices are designed not only to remove mushroom pests but to kill significant crop threats.

2) Screening of Doors and Ventilators

Mushrom flies can easily pass through ordinary wire screen and enter the mushroom house to breed on spawned compost and mushroom beds. Screening of doors and ventilators with nylon net of 35 or more than 35 mesh can effectively check the entry of flies in the cropping rooms.

3) Light Trap

Polythene sheets coated with sticky material and attached to a fluorescent strip light in each cropping room help in controlling adult flies. Insects are attracted to white light above 15 W and too yellow light at lower temperature.

4) Poison Baiting

Poison baiting with Baygon diluted with water (1:10) with addition of little sugar is an effective method of fly control in cropping rooms.

5) Disposal of Spent Compost

the spent compost and casing material contain the insects, mites and nematodes. Dumping the spent compost and csing material in moist and shady places helps it to become ideal substrate for breeding of pests. Putting this material in the compost pit and covering it with at least 10 cm thick layer of manure helps to check the fly breeding.

Chemical Methods

Treatment of compost and casing material: For treatment of compost, Lindan 20EC@20 ml per 1000 kg of straw after dilution in water is thoroughly mixed in the compost material at the last turning. If flies are observed before casing, 15ml of Lindane diluted in water is mixed well in100 Kg of ready to use casing material. Chlorpyriphos at 50 ppm in compost and at 25 ppm in casing effectively controls the larval population. However, increase in dosage level reduces the yield considerably.

Curative Method

When mushroom flies are noticed in cropping roomed, spray 30 ml Nuvan 76 EC at 22.5 g a.i./100m^3 by ULV. After spraying close the doors and ventilators for 2 hours. Avoid direct spray on mushroom beds. Observe the interval of 48 hours between spraying and picking of mushroom.

1. Flies can be killed by application of Permethrin dust (10 g a.i./kg), without any residue problem.
2. Malathion (2-3 g/m^2) can be applied between flushes and near harvest.
3. Pyrethrum at 1cc/cm applied as smoke or aerosol gives effective control of larvae and adults.
4. Spray application of Fenvalerate and Decis (0.05 per cent) on walls, floors and galleries effectively checks the adults. However, continuous application of the same chemical results in the development of resistance in insects. Therefore, care should be taken to rotate the chemicals.
5. Diflubenzuron (0.1 per cent) can be safely used between the flushes without any residue problems.

Biological Control

Basically biological control aims at pest control by the use of parasites, predator and pathogens.

I. Predators

Predators are organisms that consume their prey either partially or entirely. The typical insect predators include midges, beetles, bugs and mites. For mushroom insect pests, mites hold the greatest promise. Mites are known to be predacious on Diptera, Acarina and Nematoda.

II. Parasites

Parasites normally develop within their host over a period of time before eventually killing it. The main group of interest is Namatoda, which invade their host and are classified ad parasites. Three nematode genera namely, *Howardula, Steinernema* and *Heterorhabditis* have been reported parasitic on a large number of insect pests. In mushroom pest control, most researches have been done on *S. feltiae* and *H. bacteriophora* which have been shown parasitic on sciarids, phorids and cecids. *S. carpocapsae* alone can infect more than 250 species of insect products have been developed and some of them are being used by mushroom growers to control pest. Application of strains Sc and Sn of *Steinernema feltiae* (0.5 x 1.0 x 10^6 infective juveniles/m^2 cropping area) effectively checks the larval population.

8.5 Mites

Cultivated mushrooms are infested by several groups of mites. Fifty four species of mites have been reported from various parts of world of which 16 species have been found economically important. Cultivated mushrooms are generally infested with mites belonging to Acaridae, Pyemotidae, Ascidae, Digamasellidae, Tydeidae and Macrochelidae.

The initial infestation of mites in mushroom houses comes through raw material used for the preparation of mushroom beds. *T. dimidiatus*, a major pest of mushroom throughout the world is often present in large numbers in

Mite Infested Mushroom

hay, straw, grains and similar materials used for preparation of mushroom beds. Occasionally phorid flies also transport mushroom mites.

Symptoms/damage caused by the mites varies with the species. *T. dimidiatus* hollows out tiny buttons while in large mushrooms it makes cavities of various sizes on stalk and caps. *T. berlesei*, *T. mycophagus* and *T. longior* make holes on caps. *Tyrophagus putrescentiae* feeds on mycelium and sporophore resulting in small irregular pits on stalk and caps. *Pygmephorus* sp. feeds on mycelium below the casing layer.

Management of Mites

1. Proper pasteurization of compost and casing material.
2. Proper hygien and sanitation.
3. Disinfection of mushroom house by spraying 0.1 per cent dicofol.
4. Burning sulphur in the empty rooms @ 2-3 lb/1000 cu.ft.

5. Sterilization of empty tray
6. Applying Propargite (Omite 590 EC) at 0.888 and 0.66 g a.i./m^2 at spawning.
7. Spraying beds with chlorfenvinphos, fenitrothion, fenthion or metasystox (1.0 g a.i/m^2) immediately after spawning and before casing give satisfactory control of mites without crop reduction.
8. Spraying diazinon 20 EC (1.5-2.0 ml/10 lit. of water) in the compost or Dicofol (0.1 per cent) in mushroom beds, or drenching mushroom houses and all their premises with diazinon is also effective.

8.6 Nematode

During different phases of button mushroom production, a large number of nematodes are encountered along with insect-pests and mites. Mycelium of the fungi favourable source of food for nematodes. Nematodes are one of the most dangerous pests of button mushroom which

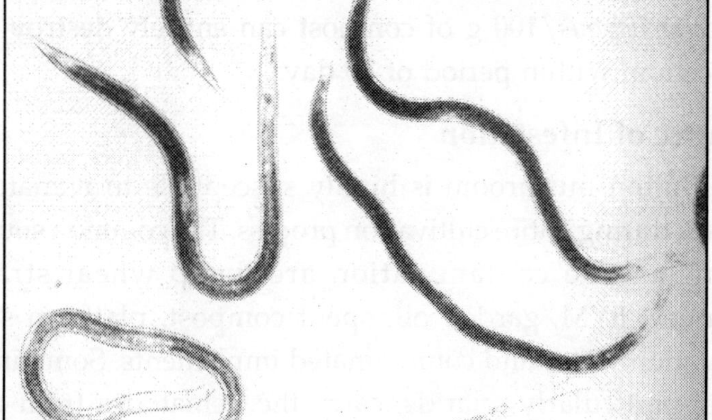

once enter the beds cannot be eradicated completely, until and unless crop beds are destroyed and disposed off completely. A number of instances of crop failure due to nematodes have been reported during the last two decades. Lack of pasteurization facility, growing crop seasonally under unhygienic conditions and lack of awareness about cultivation practices are some of the factors responsible for easy access of nematodes in cropping beds. Generally three types of nematodes *viz.* myceliophagous, saprophytic and predatory are encountered in mushroom.

Nature of Damage

Myceliophagous nematodes have needle like structure (stylet) in their mouth parts. The stylet is hollow inside and can be moved forward and backward by the contraction and relaxation of the muscles. The nematode secretes strong enzymes. These enzymes act immediately after ejection and help in penetration of stipe and to convert the cell contents in assumable forms. This nematode have very fast multiplication rate (50-100 fold/week). It has been found that initial infestation with 3 nematodes of *D. myceliophagous*/100 g of compost can entirely destroy the mycelium within period of 70 days.

Source of Infestation

Button mushroom is highly susceptible to nematode attack during entire cultivation process. The common source of nematode contamination are damp wheat straw, manure, FYM, garden oil, spent compost, platform soil, irrigation water and contaminated implements. Sometimes flies, particularly sciarids, carry the nematodes from one bed to another.

Symptoms

Following symptoms of nematode attack appear in infected beds in succession.

1. Mycelial growth is sparse, patchy and mycelium turns stingy.
2. The compost surface sinks
3. Whiteness of spawn-run slowly changes to brown.
4. Sporophore flushes are poor and delayed.
5. Browning pin head.
6. Declined yield.
7. Complete crop failure.

Management of Nematodes

Prophylactic Measures

1. Cropping should be done in purposely built mushroom houses with proper ventilation.
2. Strict hygienic and sanitation measures should be followed throughout the cropping period.
3. Composting yard must be cemented to prevent the direct contact of compost with the soil.
4. All the instruments, walls, floors and galleries should be disinfected with 4 per cent formalin.
5. Composting ingredients should always be stored in clean area.
6. Cleanliness should be maintained inside and in surrounding of mushroom farms.
7. Casing mixture should be properly pasteurized.
8. Manures used for composting should be thoroughly broken and its layers allowed to decay properly.

9. In long method of composting, covering compost with double PVC sheet for 24 hours after third turning gives effective control of nematodes.
10. Foot dips must be installed in front of each cropping room.

Chemical Control

Mushrooms, being indoor crop, provide little scope for pesticidal usage. Short life of the crop and deleterious effect of pesticides on mycelium and residue problems further limits the scope of pesticide application.

Dichlorvos (0.04 per cent) under polythene cover for 144 hours was found to be most effective for control of *A. composticola* and *Rhabditis* sp. Fenamiphos EC at 20 mg/kg incorporated in compost is practical preventive measure. Thionazin @ 80 ppm in compost and spraying on the bed surface during spawn run effectively controls the nematodes without any detectable residue on mushroom.

Physical Control

Use of heat is the most successful method of nematode control in mushroom cultivation. It is recommended that that for making compost free of nematodes, air and bed temperature in the pasteurization room must be maintained 60°C at least for two hours and cook out of mushroom house at 70°C for 5-6 hours is necessary. Dips of the appliances in boiling water for 1-2 minutes are sufficient for complete destruction of nematodes.

Chapter 9
Fungal Diseases of Mushrooms

Like all other crops, mushrooms are also affected adversely by a large number of biotic and abiotic agents/ factors. Among the biotic agents, fungi, bacteria, viruses, nematodes, insects and mites cause damage to mushrooms directly or indirectly. A number of harmful fungi are encountered in compost and casing soil during the cultivation of white button mushroom. Many of these act as competitor moulds thereby adversely affecting spawn run whereas others attack the fruit bodies at various stages of crop growth producing distinct disease symptoms. At times there is complete crop failure depending upon the stage of infection, quality of compost and environmental conditions. General distribution of various competitor moulds and pathogenic fungi is as follows:

1. *Those occurring mainly in compost include*: Olive green mould (*Chaetomium olivaceum* and other

spp.), Ink caps (*Coprinus* spp.) Green moulds (*Aspergillus* spp. *Penicillium* spp. and *Trichoderma* spp.), Black moulds (*Mucor* spp., *Rhizopus*spp.) and other (*Myriococcum praecox, Sporotrichum* sp., *Sepedonium* sp., *Fusarium* spp., *Cephalosporium* spp., *Gliocaldium* spp., and *Papulospora* spp.).

2. *Fungi occurring in compost and in casing soil*: White plaster mould (*Scopulariopsis fimicola*): Brown plaster mould (*Papulospora byssina*), Lipstick mould (*Sporendonema purpurescens*), False truffle (*Diehliomyces microsporus*) and green moulds.
3. *Fungi occurring on and in casing soil and/or on the growing mushrooms*: Cinnamon mould (*Peziza ostracoderma*), wet bubble (*Mycogone perniciosa*), Dry bubble (*Verticillium fungicola*), Cobweb (*Cladobotryum dendroides*), Pink mould (*Trichothecium roseum*) and green moulds.
4. *Fungi attacking the fruit bodies only*: Fusarial rot (*Fusarium* spp.). At any phase of growth an undesirable growth or development of certain moulds can occur and can adversely affect the final mushroom yield.

9.1 Dry Bubble

Pathogen: *Verticillium fungicola*

Common Name: *Verticillium*

This is the most common and serious fungal disease of mushroom crop. If it is left uncontrolled, disease can totally destroy a crop in 2-3 weeks (Fletcher *et al.* 1986). *Verticillium fungicola* was major pathogen responsible for considerable yield losses of cultivated mushrooms. In a disease survey of commercial mushroom houses, *V.malthousei* was isolated from 11.3 per cent of mushroom sampled (Foree *et al.* 1974). From India the first report of the heavy incidence of dry bubble disease was from mushroom farms located at Chail and Taradevi (Seth *et al.* 1973). The pathogen has been invariably isolated from the compost and casing samples collected from mushroom farms in Haryana, HP and Punjab (Sharma, 1992). Artificial inoculation with the pathogen at the time of spawning and at different loads of inoculums had delayed pinhead formation by 5 days and reduced the number and weight of fruit bodies by 2.26-47.2 per cent and 2.19-38.01 per cent, respectively (Sharma and Vijay, 1993).

Symptoms

Whitish mycelial growth is initially noticed on the casing soil which has a tendency to turn greyish yellow. If infection takes place in an early stage, typical onion shaped mushrooms are produced. Sometimes they appear as small undifferentiated masses of tissue upto 2cm in diameter. When affected at later stage, crooked and deformed mushrooms with distorted stipes and with tilted cap can be seen.When a part of the cap is affected harelip symptom is noticed. Affected mushrooms are greyish in colour. If the infection occurs at later stage, grey mouldy fuzz can be seen on the mushrooms. Sometimes little pustules or lumps appear on the cap. On fully developed sporophores, it produces localized light brown depressed spots. Adjacent spotscoalesce and form irregular brown blotches. Diseased caps shrink in blotched area, turn leathery, dry and show cracks. Infected fruit bodies are malformed, onion shaped and become irregular and swollen mass of dry leathry tissue

(Sharma, 1994). In *A.bitorquis*, the dark brown blotches caused by *V.fungicola* var *aleophilum* are sometimes covered with a layer of grey coloured mycelium particularly in the centre. In *A.bisporus* it causes minor spotting though in variety Les Miz- 60 it causes fruit body deformation. An isolate of *V.psalliote* from *A.bitorquis* causes more confluent brown spots on *A.bitorquis* but could not infect *A.bisporus* (Zaayan and Gams, 1982).

Management

a) Physical Methods

Use of sterilized casing soil, proper disposal of spent compost and proper hygiene and sanitation are essential to avoid primary infection. Treating mineral soil with aerated steam at 54.4°C for 15 minutes eliminated *V.malthousi* that had been experimently established for 17 days in axenic soil culture. Heat treatment of infected casing layer at 63°C for one hour completely prevented spore germination (Poppe, 1967).

b) Biological Method

Spraying casing soil with 100x10^6 *Trichoderma* propagules/litre/m^2 controlled *V.malthousei* in several trials on naturally infected mushroom holdings where dry bubble disease was endemic. Leaf extracts of *Callistemon lanceolatus, Cannabis sativus, Citrus* sp., *Euclyptus* sp., *Dhatura* sp., *Urtica dioica, Solanum khasianum* and *Thooja compacta* also effective for controlling the disease.

c) Chemical Methods

In laboratory trials *V.malthousei* was controlled by Zineb on a large scale, Bercema - Zineb 80 used at 0.1 - 1.2 per cent controlled the disease when used before and between the flushes (Philipp, 1963). *V.malthousei* was controlled by 3 sprays with Dithane Z-78 at 0.25 or 0.50 per cent or Hexathane at 0.30 per cent given at the time of casing, at pinhead formation and after flushes of crop. Application of chlorothalonil as a drench reduced the incidence of *V.fungicola* tolerant to certain benzimidazole fungicides. However, incorporation of chlorothalonil into the casing layer caused toxicity to crop and depressed the yield. Disease can be controlled by spraying with carbendazim, benomyl or thiophenate methyl at 100, 150 and 200g/100m2, respectively in 100-150 litres of water immediately after casing. Cased beds can also be treated with 0.5 per cent formalin or 100g carbendazim, 150g benomyl or 200g thiophenate methyl in 100-150 litres of water per m2 of bed. Fungicides triadimefon (1g/m2), prochloraz (1g/m2), Delsene M (carbendazim + maneb) (8g/m2) and chlorothalonil (2g/m2) applied after casing increased the yield from 39.9 per cent in untreated controls to 56.7, 56.3, 54.6 and 53.1 per cent, respectively.

9.2 Wet Bubble

Pathogen: *Mycogone perniciosa*

Common Name: Wet bubble, white mould, bubble, *Mycogone* disease

Wet bubble in white button mushroom incited by *Mycogone perniciosa* Magn. has been reported as one of the serious diseases from almost all the major mushroom growing countries of the world. The disease has also been reported to assume serious proportions in other major mushroom growing countries of the world such as United Kingdom, Netherlands, USA, China, Taiwan, South Africa, Brazil, Hungary, Australia and Poland from time to time. In India, this disease was reported for the first time in 1978 from some mushroom farms in Jammu and Kashmir (Kaul *et al.*, 1978). Later, this disease has been reported from the States of Himachal Pradesh, Haryana and Maharashtra (Sharma, 1994, Sharma and Kumar, 2000, Bhatt and Singh, 2000).

Symptoms

Two main symptom of disease are infected sporophores and sclerodermoid masses, which he considered to be the result of infection by *M. perniciosa* at different stages in the development of the sporophores. Thus, when infection took place before the differentiation of stipe and pileus the selerodermoid form resulted, whereas, infection after differetration resulted in the production of thickned stipe with deformation of the gills (Fletcher and Ganney, 1968). The symptoms in the form of white mouldy growth on the mushrooms, leading to their putrifaction (giving foul odour) with a golden brown liquid exudate. the sporophores may be recognised by two symptoms, one is tumorous form, infected from pinheads, and other is malformation, infected at later stage. Both types of infections may exude water drops on the surface of infected sporophores. These water drops later change into amber colour. Sharma and Kumar (2000) described the symptoms as short, curly, pure white fluffy mouldy growth of the pathogen on malformed mushrooms, which can be easily observed by nacked eyes. Cross section of deformed sporophores without cottony growth showed black circular area just beneath the upper layer. Umar *et al.* (2000) described dramatic cytological changes as a result of infection when young (up to 6mm) pin heads were infected. Large, very irregular, nodular and tumorous fungal masses are formed and no differentiation or organogenesis of the cell mass takes place.

significant quantitative and qualitative losses in the yield of *Agaricus bisporus, Pleurotus* spp., *Auricularia, Calocybe indica* and *Lentinula edodes.* Different species of *Trichoderma* which have been reported as competitors and/or pathogenic on button mushroom include, *T. viride, T. koningii, T. hamatum, T. harzianum, T. atroviride, T. pseudokoningii, T. logibrachiatum.* Among all these species, *T. harzianum* is recognised as causing the most severe problems. In India, *T. viride* was first reported by Thapa and Seth (1977) on *A. bisporus,* and *T. hamatum* and *T. harzianum* by Seth and Bhardwaj (1986-87).

Symptoms

Different species of *Trichoderma* have been reported to be associated with green mould symptoms in compost, on casing soil, in the spawn bottles and on grains after spawning. A dense, pure white growth of mycelium may appear on casing surface or in compost which resembles to mushroom mycelium. Later on mycelial mat turns to green colour may cover the entire cap surface under congenial conditions.

Control

Green moulds can be prevented by:

1. Very good hygiene
2. Proper pasteurization and conditioning of compost.
3. Sterilizing the supplements before use and mixing them throughly preferably after spawning.
4. Using the correct concentration of formalin (maximum 2 per cent)
5. Weekly sprays of mancozeb (0.2 per cent) or bavistin (0.1 per cent) TBZ (0.2 per cent) or treatment with zineb dust or Calcium hypochlorite (15 per cent) have given effective control of the disease.

9.5 False Truffle

Pathogen: ***Diehliomyces microsporus***

Common name: Truffle disease

Symptoms

The coulour of the fluffy mycelium is white to start with and turns a creamy yellow at a later stage. It appears as small wefts of white cream coloured mycelium in compost and casing soil, usually more conspicuous in the layer where compost and casing mixture meet and also on casing. Gradually the mycelial growth become thicker and develops into whitish, solid, wrinkled, rounded to irregular fungal masses resembling small brains (ascocarps of the fungus),

Symptoms

The earliest signs of the fungus consist of an inconspicuous greyish-white fine mycelium in the compost or a fine aerial growth on the compost surface 10 days after spawning. Frequently initial spawn growth is delayed and reduced. By late spawn run, fruiting structures that look like gray-green cockle-burns-1/16 inch in diameter, develop on straw in isolated spots of the affected compost. The compost will have a musty odour. Compost not supporting spawn growth generally supports the growth of *Chaetomium* and other several moulds and hence olive green mould is not the exclusive colonizer of black compost. Spawn usually grows into areas occupied by *Chaetomium,* although normal spawn growth is delayed. *C. globosum* is also noticed on spawn bottles.

Control

1. The fermentation period of the compost should not be too short. It is essential to provide an active

compost that is not too wet and has a good structure.

2. Do not add nitrogen, ammonium sulphate, urea, chicken manure or similar materials just before filling.
3. There should be sufficient time for peak-heating and sufficient supply of fresh air during pasteurization. Higher temperatures (above 60°C) for longer time should be avoided.
4. Large number of fungicides including Benomyl, Thiophanate methyl, TBZ, Vitavax, Dithane Z-78, Dithane M-45, Thiram and Captan have been found effective and sprays of Dithane Z-78 (0.2 per cent) have been recommended for checking the secondary spread.

9.7 Brown Plaster Mould

Pathogen: ***Papulaspora byssina*** **Hots.**

Symptoms

It is first noticed as whitish mycelial growth on the exposed surface of compost and casing soil in trays as well as on sides in bags due to moisture condensation. This develops further into large dense patches gradually changing colour through shades of tan, light brown to cinnamon brown; ultimately becoming rust coloured. No mushroom mycelium grows on places where plaster mould occurs.

Control

Composting should be carried out carefully, using sufficient gypsum and not too much water. Peak heating

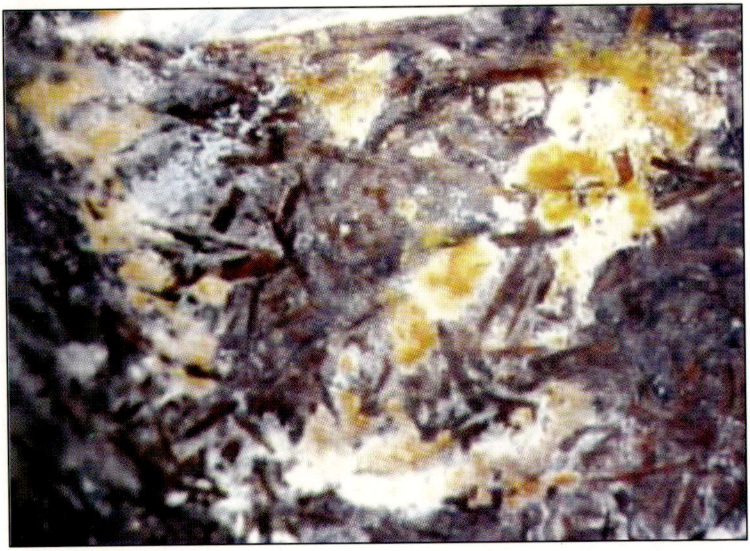

should be of sufficient duration and at proper temperatures. The compost should not be too wet before or after peak heating. Munjal and Seth (1974) recommended localized treatment of infected patches with 2 per cent formalin while Seth and Shandilya (1978) recommended 4 per cent formalin for its control. Large number of fungicides namely, benomyl, carbendazim, thiophanate methyl, vitavax, daconil, MBC, dithane Z-78, dithane M-45, captan, thiram and copper fungicides have been effective for controlling the disease. Spraying of systemic fungicides at 0.1 per cent concentration has also been recommended.

9.8 Yellow Mould (Mat disease; Vert-de. gris)

Pathogen: ***Myceliophthora lutea, Chrysosporium luteum, C. sulphureum***

Symptoms

The yellow moulds may develop in a layer below the casing (Mat disease), form circular colonies in the compost (confetti) or they may be distributed throughout the compost (Vert-de-girs). In India, *M. lutea* has been reported to induce mat disease. This fungus forms a yellow brown corky mycelial layer at the interphase of compost and casing which is difficult to detect during the impregnation of casing layer by the spawn and even during the first break. It becomes apparent when it develops its stroma like morphology and mushroom production is severely inhibited.

Control

1. Proper pasteurization of the casing mixture is very essential. Fungus does not survive the exposure for 6 hrs. at 51°C or 4 rs at 54°C.
2. Benomyl (400-500ppm) and blitox (400ppm) sprays have been found effective to control the disease and increase the yield. Spraying with calcium hypochlorite solution (15 per cent) is effective for eradication of the mould growth (Sohi, 1986).

9.9 Ink Caps

Common names: Ink weed, wild mushrooms

The appearance of inky caps during spawn run is commonly observed on the mushroom beds in northern India (Kaul *et al.*, 1978; Garcha 1984; Sohi, 1986). Artificial inculations of *C. fimetarius* at different loads of inoculum at spawning has resulted in 20.14-94.4 per cent reduction in the number of fruit bodies and 14.68 to 94.43 per cent reduction in the weight of fruit bodies (Sharma 1992).

Symptoms

Ink caps appear in the compost during spawn run or

newly cased beds and outside the manure piles during fermentation. They are slender, bell-shaped mushrooms. Cream coloured at first, blueish-black later and are usually covered with scales. This fungus sometimes grows in clusters in beds and has a long sturdy stem which often reaches deep into the compost layer. Several days after their appearance ink caps decay and form a blackish slimy mass due to autodigestion.

Control

Use properly pasteurized compost and casing soil. Avoid excessive watering. Rogue out young fruit bodies of the weed fungus to avoid its further spread.

9.10 Cinnamon Mould

Pathogen: *Chromelosporium fulva, C. ollare*

Common name: Cinnamon brown mould, brown mould

Symptoms

Although *Chromelosporium fulva (Ostracoderma fulva)* has been called cinnamon brown mould, its colour ranges from yellow gold to golden brown to cinnamon brown. The mould first appears as large circular patches of white aerial mycelium on the compost or casing. Within few days the spores are formed and the colour changes from white to light yellow or to light golden brown. As the spores mature, the colour changes to golden brown or cinnamon and the colony develops a granular appearance. The fungus produces numerous cup-like fleshy fruit bodies on beds.

Control

Casing soil should not be made completely sterile by steam or formaldehyde. Newly cased beds should be sprayed with dithane Z-78 and maintain proper moisture content in casing layer.

9.11 Lipstick Mould

Causal organism: ***Sporendonema purpurescens***

Symptoms

The disease first appears in spawned compost as a white crystalline-like mould, rather nondiscernable from spawn. As the spore of the mould mature, the colour changes from white to pink, to cherry red and then to dull orange or buff. White mycelial growth is more in loose areas of casing and can colonize well conditioned compost. In crops where there is a serious virus disease, lipstick mould usually occurs as a secondary disease.

Control

Good hygiene is essential. Good pasteurization and conditioning of the compost will eliminate the pathogen.

Chaper 10
Bacterial Diseases

10.1 Bacterial Blotch

Bacterial blotch of mushrooms is also known as brown blotch and bacterial spot.

Symptoms

Bacterial blotch of white button mushroom is characterized by brown spots or blotches on the pilei and in more severe cases, on the stipes. Circular or irregular yellowish spots develop on or near the margins of the cap which enlarges rapidly under favourable conditions and coalesce to form rich chocolate brown blotches that are slightly depressed. The most characteristic symptom of bacterial blotch is the occurrence of dark brown areas of blotches on the surface of the cap. These may be initially light in colour but may eventually become dark brown. Severely affected mushrooms may be distorted and the caps may split where the blotch symptoms occur. Brown and slightly concaved spots appeared on the surface of the diseased fruit bodies. Light infection of mushroom caps produced a yellow light brown spotting on the surface, but the common symptom associated with infection was appearance of brown, slightly sunken lesions of variable size and mushroom tissues were usually affected to a depth of 1 to 3 mm. Mushrooms often become infected at a very early stage in their development. The enlargement of the spots on the cap surface is dependent upon environmental conditions and is favoured by temperatures of at least 20°C together with the presence of water film.

Management

Ecological Management

Manipulation of relative humidity, temperature, air velocity and air movements are of great significance in managing the disease. Temperature above 20°C and relative humidity of more than 85 per cent should be avoided. Additional ventilation and air circulation after watering can ensure the quick drying of mushrooms. Temperature fluctuations at higher relative humidity leading of water condensation must be avoided.

Biological Management

Isolates of *P. fluorescnes* and other antagonistic bacteria have resulted in 30 to 60 per cent control of bacterial blotch. Many selective bacteriophages have also been found effective against *P. tolaasii* without any significant effect of *P. fluorescens.* Spraying the casing soil with a mixture of *P. fluorescens* and bacteriophage has resulted in more than 80-per cent control of blotch symptoms.

Chemical Management

Application of terramycin 9 mg per square foot, streptomycin (200 ppm), oxytetracycline (300 ppm), kasugamycin and kanamycin has been found effective in managing the disease.

Physical Management

Pasteurization of casing soils by steam/air mixture and short wave length irradiation have been reported effective in eliminating the bacterial pathogen but over-heating should be avoided otherwise biological vacuum will be created and successive invasion of moulds would be very high. The introduction of water retentive acrylic polymers as a component of casing soil mixture is also claimed to reduce the disease.

Bacterial Disease(s) of Oyster Mushroom

Till date, four bacterial pathogens namely, *Pseudomonas alcaligens*, *P. tolaasii*, *P. agarici* and *P. fluorescens* have been reported parasitising *Pleurotus* fruit bodies and causing considerable economic losses to the growers. Among these, *P.agarici* and *P. alcaligens* (not valid name) have been reported from India and are described as under:

10.2 Yellow Blotch

Occurrence and Losses

In India, heavy incidence of yellow blotch was reported (Jandaik *et al.*, 1993) which resulted in complete failure of crop in some of the mushroom units.

Symptoms

The disease appears as blotches of varying sizes on pilei sometimes depressed, yellow, hazel-brown, fawn or orange in colour. When the disease appears at primordial formation or pinhead stage, it affects the total group of early fruit bodies or only a part of them. Infected fruit bodies turn yellow and remain stunted. The slimy appearance of the infected fruit bodies under high relative humidity (more than 90 per cent) is a common symptom. If the relative humidity is less than 75 per cent, the blotched fruit bodies give appearance of burnt ulcers.

Management

Use of Chemicals

The regular application of chlorinated water containing 100-150 ppm of freely available chlorine (FCA) at 3 to 5 days interval help in minimizing losses due to bacterial pathogen, Use of oxytetracycline and streptocycline have also been reported.

Biological Management

Biocontrol of yellow blotch of oyster mushroom appears to offer a viable proposition, especially with the increasing awareness among consumers about the use of chemicals in mushroom units. The possibility of using bacteriophages as control agent for plant diseases caused by various bacterial pathogens including *Pseudomonads* has been reported and it may have application in mushroom industry as well.

The most common contaminants encountered during spawn preparation are species of *Aspergillus, Penicillium, Trichoderma, Cladosporium, Chaetomium, Alternaria, Mucor, Rhizopus, fusarium* and *Drechslera*. Grains are the main source of contamination. The fungal contamination can be usually recognized with the typical colours of their mycelium, spores or conidia. At times a distinctive zone can be recognized as lesion between inoculated mushroom mycelium and contamination. If the contaminants are allowed to grow, they may spoil a large number of spawn bags. If such contaminated bags are not timely removed and disinfected, it may become a perennial source of contamination. Selection of good quality grains, proper autoclaving and strict hygiene in spawn lab can reduce the contamination to a great extent. If the problem continues, Carbendazim or Thiophanate-methyle @ 0.05g/kg of boiled grains can help in reducing the losses caused by fungal contaminants.

Bacterial contamination is more difficult to detect. Some bacteria give a greasy appearance and emit a pungent or foul order smell. If the bacterial contamination is not detected in master spawn bottles, all the commercial spawn or compost prepared from them will become useless and may result into total loss of spawn or crop. Wet spot disease caused by *Bacillus* species is a common bacterial contamination problem in mushroom spawn. The disease can be managed by maintaining pH of spawn substrate near 6.0 and by incubating it between 20-25°C. If the problem continues, antibacterial compounds like neomycin, streptomycin or streptocyclin @ µg/gram of spawn can be added after boiling of wheat grains to manage the disease.

Chapter 11
General Guidelines

In order to decide the most effective measures for controlling a disease in mushroom, it is necessary to understand the size of the initial inoculum, density, the rate at which the disease develops and spreads and the time when the infection takes place.

Based on these, the following preventive and/or eradicative control measures are necessary for the management of these diseases:

- ☆ **Ecological**-by manipulations of environmental factors such as temperature, humidity and ventilation.
- ☆ **Biological**-by use of antagonistic organisms through incorporation of biocontrol agents and organic amendments.
- ☆ **Chemical**-by use of safe and minimum doses of specific fungicides, antibiotic etc.

A close relationship exists between crop management practices and some endemic disease problems like dry bubble, brown blotch and truffle. Biological agents are being increasingly tried throughout the world but with a limited application on commercial scale. Sanitation and hygienic measures are most essential to manage the disease particularly under Indian conditions although under certain situations use of chemicals is also inevitable.

Sanitation and Hygiene

Hygiene covers all the measures which are necessary to allow as little chance as possible to the pests and pathogens to survive, develop and spread. Thus hygiene and sanitation go hand in hand at all stages of growing mushrooms. Farm hygiene is the best defense a mushroom grower has against mushroom pests and diseases particularly during the present days, when use of chemicals on food crop is being discouraged.

After having gone through the details of different diseases discussed earlier we know that mushroom pathogens gain entry to a mushroom farm in a variety of ways. They can fly in, drift in on the wind and crawl in. Also they can be carried on people, on the vehicles and in the raw materials. What makes matter worse is that they are usually difficult or impossible to be seen with the naked eyes. Based on the critical observations during all the stages of mushroom production, the following steps have become a routine practice for successfully cultivating mushroom.

- ☆ The location of mushroom unit should be in such an area where effluents of chemical industries do not pollute the water and also the air is free from toxic fumes or gases.

- ☆ Floor for the preparation of compost should be cemented/tiled and covered with a roof.
- ☆ Substrates used for compost preparation should be fresh, protected from rain and mixed in exact proportion.
- ☆ Pasteurization and conditioning of the compost should be for optimum duration at right temperatures as over/under pasteurization may not produce quality compost and invite many disease problems.
- ☆ Do not allow free access of persons working in composting yards to spawning and other cleaner areas without changing the dress and foot-dip. Similarly, all machinery including tractors and fork-lift trucks should not be moved to the cleaner areas. After filling, all equipment and machinery should be thoroughly cleaned.
- ☆ Spawn should be fresh and free from all the contaminants.
- ☆ All equipments used for spawning, floor and walls of spawning area must be washed and disinfected.
- ☆ The fresh air should be filtered before it enters the growing rooms to exclude all particles of 2 micron and above.
- ☆ Casing mixture should be properly pasteurized (60-65°C for 5-6 hours).
- ☆ Casing mixture should be stored in a clean and disinfected place.

All the containers, equipments and machinery used for casing should be thoroughly washed and disinfected.

Keeping dust to a minimum and not to have dusty operations going on at the same time elsewhere on the farm is also very helpful.

- ☆ The pickers should use clean overalls and gloves. Picking should start from new or cleaner crop towards older crops.
- ☆ Waste from picking, chogs, trash, stems, unsaleable mushrooms should be carefully collected not allowing to fall on the floor, and be disposed off carefully.
- ☆ Avoid surface condensation of water on developing mushrooms.
- ☆ Add bleaching powder (150ppm) at every watering to manage bacterial disease.
- ☆ Remove heavily infected bags from the cropping rooms or treat the patches by spot application of 2 per cent formalin or 0.05 per cent Bavistin.
- ☆ Maintain optimum environmental conditions in the cropping rooms to avoid abiotic disorders.
- ☆ Control insect-pests well in time to avoid the spread of pathogen by them.
- ☆ At the end of crop, cooking out at 70ºC for 12 hours is very essential to eliminate all pests and pathogens.

Use of Chemicals

It is advisable to manage the disease in mushrooms through hygienic measures listed above. There are only a limited number of pesticides registered for use on mushrooms. This is because mushrooms themselves are fungi and most of the pathogens are also fungi thereby making the choice of fungicides very difficult. Moreover, because of short cropping cycle, residual toxicity of different chemicals is of great concern and it must be kept below the tolerance limit. Mushrooms are very sensitive to fumes, toxic gases and several chemicals. This also limits the frequent use of chemicals in mushroom industry. Equally important factor which limits the use of fungicides for the management of diseases in mushrooms is the problem of resistance. Repeated and regular applications of the same chemical greatly increase the chance of resistance. If equally effective alternate fungicides are available the problem of pesticide resistance can be minimized. On the other hand there are, unfortunately only a few pests or diseases that can be controlled satisfactorily by environmental manipulation alone. Some of the most common fungicides recommended for the control of major fungal pathogens of *A.bisporus* (Fletcher *et al.* 1986) and used in mushroom industry are:

- ☆ Benomyl (Benlate 50wp)- For control of *Dactylium, Mycogone, Trichoderma, Verticillium,* mix 240g/ $100m^2$ with casing or dissolve in water at 240g/ 200litres/$100m^2$ during first watering.
- ☆ Carbendazim (Bavistin) same as for benomyl.
- ☆ Chlorothalonil (Bravo or Repulse) - to control *Mycogone* and *Verticillium.* Apply as spray 2 week after casing and repeat not less than 2 weeks later @ 200ml in 100-200litre water/$100m^2$. Prochloroz Manganese (Sporgon)- to control *Mycogone, Verticillium, Dactylium,* give a single application of 300g/100litres/$100m^2$, 7-9 days after casing. For double application, use 113g/100litres/$100m^2$, 7-9 days after casing and repeat again between

second and third flushes. For triple application, use 57g/100litres/100m^2, 7-9 days after casing and after first and third flushes.

☆ Thiabendazole (Tecto)- to control *Dectylium, Mycogone, Verticillium,* apply at the same rate as Benomyl. Zineb(Zineb Tritoftoral)- to control *Dactylium, Mycogone,* Red *Geotrichum* and *Verticillum,* Use 7 per cent dust, at 350g/100m^2 every week after casing or 140g/100m^2 before watering. For wettable powder, 1kg/1000 litres @ 5 litre/100m^2 after casing and between flushes. For Tritoftoral, 5kg/100m^2 at 4.5m^2 between flushes.

Chapter 12

Constrains in Mushroom Cultivation

- ☆ Lack of motivation, people are not aware about the nutritional and medicinal value of mushrooms
- ☆ Use of mushrooms is only limited to five star hotels and people believe it as a luxurious item
- ☆ Mushroom being a perishable commodity availability of domestic market or processing industry is essential
- ☆ Lack of trained mycologist or trained personnel and training facilities
- ☆ The availability of spawn
- ☆ Lack of location specific standard cultivation technology for potential mushrooms

Chapter 13
Future Prospects

- ☆ There is an urgent need for the wider publicity about merits of mushrooms, convince farmers for diversification.
- ☆ Exploring the naturally occurring edible mushroom in Gujarat.
- ☆ Development of domestic market for mushrooms.
- ☆ Identifying the area for cultivation of specific mushroom species in Gujarat.
- ☆ Establishment of mushroom research, training and extension centre of excellence in Agricultural universities.
- ☆ Standardization of location specific cultivation techniques of suitable species of mushrooms.
- ☆ Involvement of public - private sector or NGOs in educating people regarding importance and cultivation of mushrooms.

Chapter 14
For Better Mushroom Industry in India

Most growers in India are self-employed and operating small-scale farms. They have different backgrounds with low or no knowledge of running small biological enterprises. Many short-sighted and non-committed growers are getting out of mushroom growing enterprises due to small setbacks they encounter before they accumulate enough experience in mushroom cultivation management. This situation creates fluctuations in the total number of oyster mushrooms growing units and causes an inconsistent supply-demand curve in the marketplace. This in turn causes the market price for oyster mushroom producing growers to fluctuate. As such, the market of oyster mushroom is highly localized with individual traders having great control on prices. The retail price of fresh oyster mushroom varies in India from INR*30-120 (USD0.66-2.65) per kg.

To make the oyster mushroom growing business more profitable, the following efforts should be made:

- The present growers must join hands to form co-operative societies in order to share the technical information on day to day growing and spawn production, and in order to control the price of mushrooms in the market place. The idea of co-operative formation among growers is already proposed among the grower communities but no leader has emerged to date.
- The costs of production should be maintained as low as possible by utilizing the local agricultural residue.
- Introduction of value-added products like oyster mushroom powder for soups, and oyster mushroom pizza should be made.
- Mature and committed entrepreneurs should be encouraged to become involved in the mushroom industry.
- The Indian government sectors must take the initiative in assisting in the marketing of fresh and processed oyster mushrooms for export by purchasing crops from the small scale mushroom farms. The revenue from this process can later be utilized in improving the rural infrastructure.

Bibliography

Bhatt, N and Singh, RP 2000. Incidence and lossess in yield by fungal pathogens encountered from the beds of *A. bisporus*. Indian. J. Mush. 18: 46-49

Chang S. T. 1974. Production of the straw mushroom *(Volvariella volvacea)* from cotton wastes. The Mushroom J. 21: 348-353.

Chang S. T. 1996. Mushroom research and development - equality and mutual benefit. In Mushroom Biology and Mushroom Products. Royse D. J., ed., University Park, Pan. 1-10.

Chang, S. T. and J. A. Buswell. 1996. Mushroom nutriceuticals. World J. Microb. and Biotech. 12:473-476.

Chang, S. T. and J. A. Buswell. 1999. Ganoderma lucidum (Curt.:Fr.) P. Karst. (Aphyllophoromycetidease): a mushrooming medicinal mushroom. *Int. J. Med. Mush.* 1: 139-146.

Chang, S. T. and P. G. Miles. 1989. Edible mushrooms and their cultivation. CRC Press, Inc. Florida, 345p.

Chang, S. T. and P. G. Miles. 2004. Mushrooms: Cultivation, Nutritional Value, Medicinal Effect, and Environmental Impact (Second Edition).CRC Press.Boca Raton, 451pp.

Chang, S. T. and W. A. Hayes (eds.) 1978. The Biology and Cultivation of Edible Mushrooms. Academic Press, Inc., New York and London. 819 pp.

Cheung, P. C. K. 1996. The hypocholesterolemic effect of two edible mushrooms: Auricularia auricula and Tremella fuciformis in hypercholesterolemic rats. Nutrition Res. 16:1721 -1725.

Chiu, K. W., A. H. W. Lam and P. K. T. Pang. 1995. Cardiovascular active substances from the straw mushroom, Volvariella volvacea. Phytotherapy Res. 9: 93-99.

Crisan, E. V. and A. Sands. 1978. Nutritional value. In "The Biology and Cultivation of Edible Mushrooms" edited to S. T. Chang and W. A. Hayes. Academic Press, New York.

Dewhurst M. 2002. Phase III -the future? The Mushroom J. 626, 17-18.

Fletcher JT, White PF and Gaze RH. 1986. Mushroom pest and Disease Control. *Intercept Ltd. London.* 149pp.

Garcha, HS, Khanna, PK and Sandhu, GS. 1987. Status of pests in the cultivated mushrooms in India. In: *Cultivating Edible Fungi* (Eds Wuest PJ, Gela Royse DJ and Beelemen RB) Elsevier Sci. Pub. The Netherlands pp 649-665

Garcha, HS. 1984. *A Manual of Mushroom Growing*. PAU publication 71p

Huang, B. H., K. H. Yung and S. T. Chang. 1985. The sterol composition of Volvariella volvacea and other edible mushrooms. Mycologia 77:959-963.

Jandaik, CL, Sharma, VP and Raina, R. 1993. Yellow blotch of *Pleurotus sajor-caju* (fr) Singer-a bacterial disease new to India. *Mush. Res.* 2: 45-48

Kaul, T. N. and B. L. Dhar. 2007. Biology and Cultivation of Edible Mushrooms. Westville Publishing House. New Delhi, 240pp

Kaul, T. N. and B. L. Dhar. 2007. Biology and Cultivation of Edible Mushrooms. Westville Publishing House. New Delhi, 240pp

Kaul, TN, Kachroo, JL and Ahmed, N. 1978. Diseases and competitors of mushroom farms in Kashmir Valley. *Indian Mush. Sci.* 1: 193-203

Lemmers G. 2003. The merits of bulk Phase III. The Mushr J. 642, 17-22.

Liu, G. T. 1999. Recent advances in research of pharmacology and clinical applications of Ganoderma P. K Karst. species (Aphyllophoromycetideae) in China. *Int. J. Med. Mush.* 1:63-67.

Mizuno, T., H. Saito, T. Nishitoba and H. Kawagishi. 1995. Antitumour- active substances from mushrooms. Food Reviews International 11:23-61.

Quimio, T. H., S. T. Chang and D. J. Royse. 1990. Technical guidelines for mushroom growing in the tropics. FAO, Rome, 155p

Seth, PK and Bhardwaj, SC. 1989. Studies on vert-de gris caused byb *Myceliophthora lutea* Coast on *A. bisporus* and its control. *Mush. Sci.* 12(2): 725-733

Seth, PK and Dar, GM. 1989. Studies on *Cladobotryum dendroides* causing cobweb disease of *A. bisporus* and its control. *Mush. Sci.* 12(2): 711-723

Seth, PK, Kumar, S and Shandilya, TR. 1973. Combating dry bubble of mushrooms. *Indian Hort.* 18(2): 17-18

Seth, PK. 1977. Pathogens and competitor of *A. bisporus* and their control. *Indian J. Mush.* 3: 31-40

Sharma SR and Vijay, B. 1996. Prevelance and interaction of competitor and parasitic moulds in *A. bisporus*. *Mush. Res.* 5(1): 13-18

Sharma, SR and Kumar, S. 2000. Studies on wet bubble disease of white button mushroom, *A. bisporus* caused by *M. perniciosa*. *Mush. Sci.* 15(2): 569-575

Sharma, SR and Kumar, S. 2000. Viral diseases of mushrooms. In: *Diseases of Horticultural Crops Vegetables, Ornamentals and Mushrooms* (Verma, LR and Sharma RC eds). Pp 166-178. Sceintific Publishers Jodhpur.

Sharma, SR and Vijay, B. 1993. Competitor moulds- a serious threat to *A. bisporus* cultivation in India. Proc. Golden jubilee Symp. Hort. Soc. India. Bangalore p 312-313

Sharma, SR. 1992. Compost and casing mycoflora from mushroom farms on northern India. *Mush. Res.* 1: 119-121

Sharma, SR. 1994. Survey for diseases in cultivated mushrooms. *Ann. Rep. NRCM,* pp23

Sharma, SR. 1994. Viruses in mushrooms. In: *Mushroom Biotechnology* (Nair MC, Gokulapalan C, Das L eds). Pp 658-685. Indus Publishing Company, New Delhi.

Sharma, VP, Sharma, SR and Kumar, S. 1999. Adverse effect of formaldehyde on some edible fungi and their mycoparasites. *Mush. Res.* 8(2): 23-26

Sharma, VP, Suman, BC, Guleria, DS. 1992. *Cladobotryum verticillium* a new pathogen of *A. bitorquis. Indian J. Mycol. Pl. Path.* 22(1): 62-65

Sohi, HS. 1986. Diseases and competitor moulds associated with mushroom culture and their control. *Extension Bull.* No. 2-12p

Tam S. C., K. P. Yip, K. P. Fung and S. T. Chang. 1986. Hypotensive and renal effects of an extract of the edible mushroom, *Pleurotus sajor-caju. Life Sciences* 38: 1155-1161.

Thapa, CD, Seth, PK and Pal, J. 1979. Occurrence of olive green mould *(Chaetomium globosum)* in mushroom beds and its control. *Indian. J. Mush.* 5: 9-13

Van Griensven, L.J.L.D.(eds.,) 1988. The Cultivation of Mushrooms. Darlington Mushroom Lab. Ltd. Sussex, England.

Wang H. X., T. B. Ng, W. K. Liu, V. E. C. Ooi and S. T. Chang. 1996b. Polysaccharide-peptide complexes from the cultured mycelia of the mushroom Coriolus versicolor and their culture medium activate mouse lymphocytes and macrophages. *Int. J. Biochem. Cell Bio.* 28: 601-607.

Wasser, P. S., E. Nevo, D. Sokolov, S. Reshetnikov, and M. Timor-Tismenetsky. 2000. Dietary supplements from medicinal mushrooms: diversity of types and variety of regulations. *Intl. J. Medi. Mushrooms* 2:1-19.

Yao, Q. Z., M. M. Yu, L. S. M. Ooi, T. B. Ng, S. T. Chang, S.S.M. Sun and V. E. C. Ooi. 1998. Isolation and characterization of a Type I ribosome-inactivating protein from fruiting bodies of the edible mushroom *(Volvariella volvacea)*. J. Agri. Food Chem. 46: 788-792.

Yau C. K. and S. T. Chang. 1972. Cotton waste for indoor cultivation of straw mushroom. World Crops 24: 302-303.

Index